Springer

New York
Berlin
Heidelberg
Hong Kong
London
Milan
Paris
Tokyo

Books by Serge Lang
of Interest for High Schools

Geometry (with Gene Murrow)

This high school text, inspired by the work and educational interests of a prominent research mathematician, and Gene Murrow's experience as a high school teacher, presents geometry in an exemplary and, to the student, accessible and attractive form. The book emphasizes both the intellectually stimulating parts of geometry and routine arguments or computations in physical and classical cases.

MATH! Encounters with High School Students

This book is a faithful record of dialogues between Lang and high school students, covering some of the topics in the *Geometry* book, and others at the same mathematical level. These encounters have been transcribed from tapes, and are thus true, authentic, and alive.

Basic Mathematics

This book provides the student with the basic mathematical background necessary for college students. It can be used as a high school text, or for a college precalculus course.

The Beauty of Doing Mathematics

Here, we have dialogues between Lang and audiences at a science museum in Paris. The audience consisted of many types of persons, including some high school students. The topics covered are treated at a level understandable by a lay public, but were selected to put people in contact with some more advanced research mathematics which could be expressed in broadly understandable terms.

First Course in Calculus

This is a standard text in calculus. There are many worked out examples and problems.

Introduction to Linear Algebra

Although this text is used often after a first course in calculus, it could also be used at an earlier level, to give an introduction to vectors and matrices, and their basic properties.

Serge Lang
Gene Murrow

Geometry

Second Edition

With 577 Illustrations

Springer

Serge Lang
Department of Mathematics
Yale University
10 Hillhouse Avenue
PO Box 208283
New Haven, CT 06520-8283
USA

Gene Murrow
Riverview Farm Road
Ossining, NY 10562
USA

ISBN 978-1-4419-3084-2

Mathematics Subject Classification (2000): 51-01, 51M05, 00-01

Library of Congress Cataloging-in-Publication Data
Lang, Serge
    Geometry / Serge Lang, Gene Murrow. — 2nd ed.
    Includes index.
    1. Geometry.   I. Murrow, Gene.   II. Title.
QA445.L36   1988   516.2   87-32376

Printed on acid-free paper.

Contents

Introduction

The present book is intended as a text for the geometry course in secondary schools. Several features distinguish it from currently available texts.

Choice of topics. We do not think that the purpose of the basic geometry course should be to do geometry *a certain way*, i.e. should one follow Euclid, should one not follow Euclid, should one do geometry the transformational way, should one do geometry without coordinates, etc.? We have tried to present the topics of geometry whichever way seems most appropriate.

The most famous organization of geometrical material was that of Euclid, who was most active around 300 B.C. Euclid assembled and enhanced the work of many mathematicians before him, like Apollonius, Hippocrates, Eudoxus. His resulting textbook, *The Elements*, was used virtually unchanged for 2,000 years, making it the most famous schoolbook in history.

Many new ideas have been added to the body of knowledge about geometry since Euclid's time. There is no reason *a priori* to avoid these ideas, as there is no reason to push them excessively if inappropriate.

For certain topics (e.g. in Chapters 1, 5, 6, 7), Euclid's way is efficient and clear. The material in Chapters 3 and 4 on Pythagoras' theorem also follows Euclid to a large extent, but here we believe that there is an opportunity to expose the student early to coordinates, which are especially important when considering distances, or making measurements as applications of the Pythagoras theorem, relating to real life situations. The use of coordinates in such a context does not affect the logical structure of Euclid's proofs for simple theorems involving basic geometric figures like triangles, rectangles, regular polygons, etc.

An additional benefit of including some sections on coordinates is that algebraic skills are maintained in a natural way throughout the year principally devoted to geometry. Coordinates also allow for practical computations not possible otherwise.

We feel that students who are subjected to a secondary school program during which each year is too highly compartmentalized (e.g. a year of geometry from which *all* algebra has disappeared) are seriously disadvantaged in their later use of mathematics.

Experienced teachers will notice at once the omissions of items traditionally included in the high school geometry course, which we regard as having little significance.

Some may say that such items are fun and interesting. Possibly. But there are topics which are equally, or even more, fun and interesting, and which in addition are of *fundamental importance*. Among these are the discussion of changes in area and volume under dilation, the proofs of the standard volume formulas, vectors, the dot product and its connection with perpendicularity, transformations. The dot product, which is rarely if ever mentioned at the high school level, deserves being included at the earliest possible stage. It provides a beautiful and basic relation between geometry and algebra, in that it can be used to interpret perpendicularity extremely efficiently in terms of coordinates. See, for instance, how Theorem 10-2 establishes the connection between the Euclidean type of symmetry and the corresponding property of the dot product for perpendicularity.

The proofs of the standard volume formulas by means of dilations and other transformations (including shearing) serve, among others, the purpose of developing the student's spatial geometric intuition in a particularly significant way. One benefit is a natural extension to higher dimensional space.

The standard transformations like rotations, reflections, and translations seem fundamental enough and pertinent enough to be mentioned. These different points of view are not antagonistic to each other. On the contrary, we believe that the present text achieves a coherence which never seems forced, and which we hope will seem most natural to students who come to study geometry for the first time.

The inclusion of these topics relates the course to the mathematics that precedes and follows. We have tried to bring out clearly all the important points which are used in subsequent mathematics, and which are usually drowned in a mass of uninteresting trivia. It is an almost universal tendency for elementary texts and elementary courses to torture topics to death. Generally speaking, we hope to induce teachers to leave well enough alone.

Proofs. We believe that most young people have a natural sense of reasoning. One of the objectives of this course, like the "standard" course, is to develop and systematize this sense by writing "proofs". We do not wish to oppose this natural sense by confronting students with an unnatural logical framework, or with an excessively formalized axiomatic system. The order which we have chosen for the topics lends itself to this attitude. Notions like distance and length, which involve numerical work, appear at the beginning. The Pythagoras theorem, which is by far the most important theorem of plane geometry, appears immediately after that. Its proof is a perfect example of the natural mixture of a purely geometric idea and an easy algebraic computation.

In line with the way mathematics is usually handled we allow ourselves and the student to use in the proofs facts from elementary algebra and logic without cataloguing such facts in a pretentious axiomatic system. As a result, we achieve clearer and shorter chains of deduction. Whatever demerits the old books had, they achieved a certain directness which we feel should not be lost.

On the other hand, we are still close to Euclid. We preferred to use a basic axiom on right triangles at first (instead of Euclid's three possibilities SSS, SAS, ASA) for a number of reasons:

It suffices for the proofs of many facts, exhibiting their symmetry better.

It emphasizes the notion of perpendicularity, and how to use it.

It avoids a discussion of "congruence".

Of course, we also state Euclid's three conditions, and deduce further facts in a standard manner, giving applications to basic geometric figures and to the study of special triangles which deserve emphasis: 45-45-90 and 30-60-90 triangles.

We then find it meaningful to deal with the general notion of congruence, stemming from mappings (transformations) which preserve distance. At this point of course, we leave the Euclidean system to consider systematically rotations, translations, and reflections. We also show how these can be used to "prove" Euclid's three conditions. In many ways, such proofs are quite natural.

Exercises. While the exercise sets include many routine and drilling problems, we have made a deliberate effort to include a large number of more interesting ones as well. In fact, several familiar (but secondary) theorems appear as exercises. If one includes *all* such theorems in the text itself, only overly technical material remains for the student to practice on and the text becomes murky. In addition, it is pedagogically sound to allow students a chance to figure out some theorems for themselves before they see the teacher do them. One cannot go too far in this direction. Even for the theorems proved at length in the text, one might tell the students to try to find the proof first by themselves, before it is done in class, or before they read the proof in the book. In some cases, students might even find an alternative proof.

This policy has some consequences in the teaching of the course. The teacher should not be afraid to spend large amounts of class time discussing interesting homework problems, or to limit some assignments to two or three such exercises, rather than the usual ten to fifteen routine ones. The students should be reassured that spending some time thinking about such an exercise, even when they are not able to solve it, is still valuable.

We feel that even if the secondary results included as exercises were for the most part entirely omitted from the course, even then the students would not be hampered in their further study of mathematics. Any subject (and especially one as old as geometry) accumulates a lot of such results over the years, and some pruning every few centuries can only be healthy.

The experiment and construction sections are especially suited to in-class activity, working in groups, and open-ended discussions, as an alternative to the daily class routine.

A Coherent Development. Like most mathematics teachers, we are aware of the controversy surrounding the geometry course, and the problem of structuring a course which is broader than the traditional Euclidean treatment, but which preserves its pedagogical virtues. We offer this book as one solution. Reforms of the curriculum cannot proceed by slogans—New Math, Old Math, Euclid Must GO, etc. We are trying to achieve reform by proposing a concrete, coherent development, not by pushing a new ideology.

SERGE LANG AND GENE MURROW
Spring 1988

Acknowledgement. We would like to thank those people who have made suggestions and pointed out misprints in the preliminary edition, especially Johann Hartl, Bruce Marshall, W. Schmidt, S. Schuster, and Samuel Smith. We are also very grateful to Philip Carlson for providing an answer manual (also Springer-Verlag).

S. L. and G. M.

CHAPTER 1

Distance and Angles

1, §1. LINES

The geometry presented in this course deals mainly with figures such as points, lines, triangles, circles, etc., which we will study in a logical way. We begin by briefly and systematically stating some basic properties.

For the moment, we will be working with figures which lie in a plane. You can think of a plane as a flat surface which extends infinitely in all directions. We can represent a plane by a piece of paper or a blackboard.

LIN. *Given two distinct points P and Q in the plane, there is one and only one line which goes through these points.*

We denote this line by L_{PQ}. We have indicated such a line in Figure 1.1. The line actually extends infinitely in both directions.

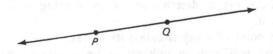

Figure 1.1

We define the **line segment**, or **segment** between P and Q, to be the set consisting of P, Q and all points on the line L_{PQ} lying between P and Q. We denote this segment by $\overline{PQ}$.

If we choose a unit of measurement (such as the inch, or centimeter, or meter, etc.) we can measure the length of this segment, which we denote $d(P, Q)$. If the segment were 5 cm long, we would write $d(P, Q) = 5$ cm. Frequently we will assume that some unit of length has been fixed, and so will write simply $d(P, Q) = 5$, omitting reference to the units.

Two points P and Q also determine two **rays**, one starting from P and the other starting from Q, as shown in Figure 1.2. Each of these rays starts at a particular point, but extends infinitely in one direction.

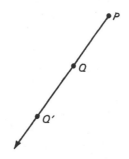

(a) Ray starting from P
    passing through P and Q

(b) Ray starting from Q
    passing through P and Q.

Figure 1.2.

Thus we define a **ray** starting from P to consist of the set of points on a line through P which lie to one side of P, and P itself. We also say that a ray is a **half line**.

The ray starting from P and passing through another point Q will be denoted by R_{PQ}. Suppose that Q' is another point on this ray, distinct from P. You can see that the ray starting from P and passing through Q is the same as the ray that starts from P and passes through Q'. Using our notation, we would write

$$R_{PQ} = R_{PQ'}.$$

In other words, a ray is determined by its starting point and by any other point on it.

The starting point of a ray is called its **vertex**.

Sometimes we will wish to talk about lines without naming specific points on them; in such cases we will just name the lines with a single letter, such as K or L. We define lines K and L to be **parallel** if either $K = L$, or $K \neq L$ and K does not intersect L. Observe that we have allowed that a line is parallel to itself. Using this definition, we can state three important properties of lines in the plane.

PAR 1. *Two lines which are not parallel meet in exactly one point.*

PAR 2. *Given a line L and a point P, there is one and only one line passing through P, parallel to L.*

In Figure 1.3(a) we have drawn a line K passing through P parallel to L. In Figure 1.3(b) we have drawn a line K which is not parallel to L, and intersects at a point Q.

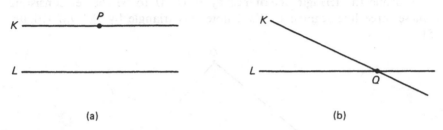

(a)   (b)

Figure 1.3

PAR 3. *Let L_1, L_2, and L_3 be three lines. If L_1 is parallel to L_2 and L_2 is parallel to L_3, then L_1 is parallel to L_3.*

This property is illustrated in Figure 1.4:

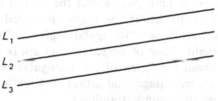

Figure 1.4

We define two segments, or two rays, to be **parallel** if the lines on which they lie are parallel.

To denote that lines L_1 and L_2 are parallel, we use the symbol

$$L_1 \parallel L_2.$$

Note that we have *assumed* properties **PAR 1, PAR 2, PAR 3**. That is, we have accepted them as facts without any further justification. Such facts are called **axioms** or **postulates**. The statement **LIN** concerning lines passing through two points was also accepted as an axiom. We discuss this more fully in §4 on proofs.

You should already have an intuitive idea about what a triangle is. Examine the following precise definition and see that it fits your intuitive idea:

Let P, Q, and M be three points in the plane, not on the same line. These points determine three line segments, namely,

$$\overline{PQ}, \qquad \overline{QM}, \qquad \overline{PM}.$$

We define the **triangle determined by** P, Q, M to be the set consisting of these three line segments. We denote this triangle by $\triangle PQM$ (Figure 1.5).

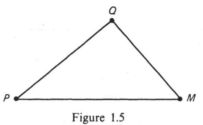

Figure 1.5

Note that if P, Q, and M *do* lie on the same line, we do not get a triangle. We define points to be **collinear** whenever they lie on the same line.

Remark. There is some ambiguity about the notion of a triangle. The word is sometimes used to denote the region bounded by the three line segments. One should really say "**triangular region**". On the other hand, we shall commit a slight abuse of language, and speak of the "area of a triangle", when we mean the "area of the triangular region bounded by a triangle". This is current usage, and although slightly incorrect, it does not really lead to serious misunderstandings.

We define a triangle to be **isosceles** if two sides of the triangle have the same length. We define the triangle to be **equilateral** if all three sides of the triangle have the same length. Pictures of such triangles are drawn on Figure 1.6(a) and (b).

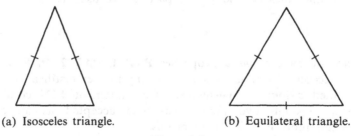

(a) Isosceles triangle.   (b) Equilateral triangle.

Figure 1.6.

Remark on "Equality". In mathematics, two quantities are said to be **equal** if they are **the same.** For instance, the two segments $\overline{AB}$ and $\overline{CD}$ in Figure 1.7(a) and (b) are *not* equal. They are *not the same* segment. They constitute *two* segments.

(a)   (b)

Figure 1.7

However, their **lengths** are **equal.** Similarly, the two angles of Figure 1.8(a), (b) are not equal, but their **measures** are **equal.**

(a)   (b)

Figure 1.8

If we say that a point P equals a point Q, then P and Q are *the same point*, merely denoted by two different letters. (The same person may have two different names: a first name and a last name!)

1, §1. EXERCISES

1. Illustrate each of the following by labeling two points on your paper P and Q and drawing the picture.
   (a) L_{PQ} (b) R_{QP} (c) $\overline{PQ}$ (d) R_{PQ}

2. Draw three points M, P, and Q, and draw the rays R_{PM} and R_{PQ}. Under what conditions will these two rays together form a whole line?

3. Suppose points P, Q M lie on the same line as illustrated in Figure 1.9. Write an equation relating $d(P, Q)$, $d(P, M)$, and $d(Q, M)$.

          P   Q   M

Figure 1.9

4. By our definitions, is $\overline{AB}$ parallel to $\overline{PQ}$?

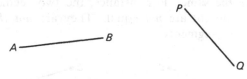

Figure 1.10

5. Criticize the following statement:

   The line segments connecting any three points in the plane make up a triangle.

6. In Figure 1.11, line K is parallel to line U, and line L intersects line K at point P. What can you conclude about lines L and U? Why?

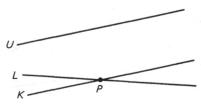

Figure 1.11

7. In Figure 1.3(a) on page 3, we have drawn a picture illustrating **PAR 2**, where point P is not on line L. Explain how **PAR 2** is still true even if P does lie on line L.

CONSTRUCTION 1-1

Since we study figures in the plane, it is useful to know how to draw accurate pictures of them. We sometimes use these pictures to make guesses about properties of various figures. Below we introduce the first of some basic constructions. In all constructions, work carefully, use a sharp pencil, and don't be afraid to try different things on your own.

To construct a triangle given the lengths of the three sides.

You are given segments $\overline{AB}$, $\overline{CD}$, and $\overline{EF}$ (Figure 1.12), and you wish to construct a triangle whose sides have lengths equal to the lengths of the three segments.

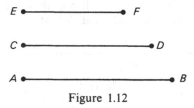

Figure 1.12

Draw a line on your paper. Set your compass at a distance equal to the length of $\overline{AB}$. Choose a point P on your line, place the point of the compass (the tip with the needle) on P, and draw an arc which crosses the line at a point Q. Set the compass tips at a distance equal to the length of $\overline{CD}$, place a tip on P, and draw an arc above the line. Set the compass equal to the length of $\overline{EF}$, place a tip on Q, and draw another arc crossing the previous one. Where these arcs intersect is point R, and triangle PQR is the desired one (Figure 1.13).

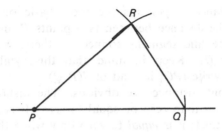

Figure 1.13

EXPERIMENT 1-1

1. Construct a triangle with sides equal in length to $\overline{PQ}$, $\overline{RS}$, and $\overline{XY}$ (Figure 1.14).

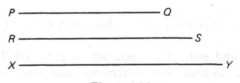

Figure 1.14

2. Construct a triangle whose sides are 5 cm, 7 cm, and 10 cm.

3. Choose one of the segments in Problem 1, and construct an equilateral triangle, whose sides have the same length as that segment.

4. Construct a triangle such that one of the sides has the same length as $\overline{PQ}$, while the two other sides have the same length as $\overline{RS}$.

5. Construct a triangle whose sides have lengths 5 cm, 7 cm, and 15 cm.

6. Can you explain why there is trouble with Problem 5?

7. Draw any triangle just using a ruler, and measure the length of the three sides. Add up the lengths of any two sides and compare this total with the length of the third side. What do you notice?

8. Construct a triangle with sides of length 5 cm, 10 cm, and 15 cm. What happens?

9. Let P, Q, and M be three points in the plane, and suppose

$$d(P, Q) + d(Q, M) = d(P, M).$$

What can you conclude about point P, Q, and M? Draw a picture.

1, §2. DISTANCE

The notion of distance is perhaps the most basic one concerning the plane. We define the **distance** between two points P and Q in the plane as the length of the line segment connecting them, which we have already denoted $d(P, Q)$. Keep in mind that this symbol stands for a number. We often write $|PQ|$ instead of $d(P, Q)$.

A few ideas about distance are obvious. The distance between two points is either greater than zero or equal to zero. It is greater than zero if the points are distinct; it is *equal* to zero only when the two points are in fact the same—in other words when they are *not* distinct. In addition, the distance from a point P to a point Q is the same as the distance from Q back to P.

We write these properties of distance using proper symbols as follows:

DIST 1. *For any points P, Q, we have $d(P, Q) \geqq 0$. Furthermore, $d(P, Q) = 0$ if and only if $P = Q$.*

DIST 2. *For any points P, Q, we have $d(P, Q) = d(Q, P)$.*

The phrase "if and only if" which we used in **DIST 1** is the way a mathematician condenses two statements into one. In particular, we could replace that phrase in **DIST 1** with the two statements:

$$\text{If } P = Q, \text{ then } d(P, Q) = 0$$

and

$$\text{If } d(P, Q) = 0, \text{ then } P = Q.$$

Study **DIST 1** and **DIST 2** carefully and convince yourself that they express the ideas mentioned in the previous paragraph and that they fit your intuition. We discuss "if and only if" some more in Experiment 1-2, and in a later section when we talk about proof.

Another important property of distance is one you might have discovered for yourself in Experiment 1-1. It is called the

Triangle Inequality. *Let P, Q, M be points. Then*

$$d(P, M) \leqq d(P, Q) + d(Q, M).$$

In the case that $d(P, M) < d(P, Q) + d(Q, M)$, points P, Q, and M determine a triangle illustrated in Figure 1.15.

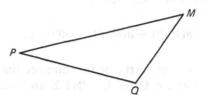

Figure 1.15

This statement tells us that the sum of the lengths of any two sides of a triangle is greater than the length of the third side. Notice that the triangle inequality allows the case where $d(P, Q) + d(Q, M) = d(P, M)$. Under what circumstances does this happen? The answer (which you might also have found in Experiment 1-1) is given by the following property.

SEG. *Let P, Q, M be points. We have*

$$d(P, Q) + d(Q, M) = d(P, M)$$

if and only if Q lies on the segment between P and M.

This property **SEG** certainly fits our intuition of line segments, and is illustrated in Figure 1.16(a) where Q lies on the segment $\overline{PM}$; and in (b) and (c) where Q does not lie on this segment.

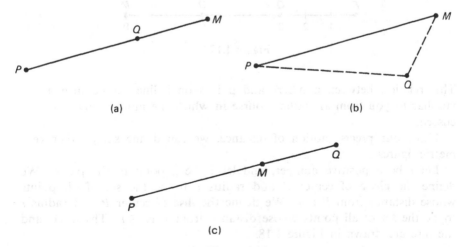

(a)

(b)

(c)

Figure 1.16

Again, the "if and only if" is saying *two* things. The first is:

If Q lies on the segment $\overline{PM}$, then

$$d(P, Q) + d(Q, M) = d(P, M).$$

This is really just a basic property of the number line which we can use as a fact, just as we will use **DIST 1**, **DIST 2**, and the **triangle inequality**. The second is: If

$$d(P, Q) + d(Q, M) = d(P, M),$$

then Q lies on $\overline{PM}$. This is not so obvious, but we have already verified it somewhat in the previous Experiment.

There is one final property concerning distance and segments which we will use. The points on a segment $\overline{PM}$ can be described by all numbers between 0 and $d(P, M)$. For example, suppose $d(P, M) = 9$. If we choose any number c such that

$$0 \leqq c \leqq 9,$$

there is just one point on the segment whose distance from P is equal to c. Another example is a ruler. Each *number* on the ruler corresponds to a point a certain distance from one end. In Figure 1.17 we have drawn a point Q on the segment whose distance from P is 3, and a point Q' whose distance from P is $2 \cdot 3 = 6$.

Figure 1.17

This relation between numbers and points on a line should already be familiar to you from an earlier course in which the number line was discussed.

Using our precise notion of distance, we can define some other geometric figures.

Let r be a positive number, and let P be a point in the plane. We define the **circle** of center P and radius r to be the set of all points whose distance from P is r. We define the **disc** of center P and radius r to be the set of all points whose distance from P is $\leqq r$. The circle and the disc are drawn in Figure 1.18.

Figure 1.18

Though we have only discussed distance between two points, we intuitively define the **circumference** of a circle as the distance "once around"
the circle. This idea is more fully developed in §4 of Chapter 8.

1, §2. EXERCISES

1. Radio station KIDS broadcasts with sufficient strength so that any town 100
   kilometers or less but no further from the station's antenna can receive the
   signal.
   (a) If the towns of Ygleph and Zyzzx pick up KIDS, what can you conclude
       about their distances from the antenna?
   (b) If a messenger were to travel from Ygleph to the antenna and then on to
       Zyzzx, he would have to travel at most how many kilometers?
   (c) What is the maximum possible distance between Ygleph and Zyzzx? Ex-
       plain why your answer is correct.

2. Charts indicate that city B is 265 km northwest of city A, and city C is
   286 km southwest of city B. What can you conclude about the distance from
   city A directly to city C?

3. Which of the following sets of lengths could be the lengths of the sides of a
   triangle:
   (a) 2 cm, 2 cm, 2 cm   (b) 3 m, 4 m, 5 m   (c) 5 cm, 8 cm, 2 cm
   (d) 3 km, 3 km, 2 km   (e) $1\frac{1}{2}$ m, 5 m, $3\frac{1}{2}$ m   (f) $2\frac{1}{2}$ cm, $3\frac{1}{2}$ cm, $4\frac{1}{2}$ cm

4. If two sides of a triangle are 12 cm and 20 cm, the third side must be larger
   than _____ cm, and smaller than _____ cm.

5. Let P and Q be distinct points in the plane. If the circle of radius r_1 around
   P intersects the circle of radius r_2 around Q in *two* points, what must be true
   of $d(P, Q)$?

6. If $d(X, Y) = 5$, $d(X, Z) = 1\frac{1}{2}$, and Z lies on $\overline{XY}$, then $d(Z, Y) = ?$.

7. Draw a line segment $\overline{AB}$ whose length is 15 cm. Locate points on $\overline{AB}$ whose
   distances from A are:
   (a) 3 cm;   (b) $\frac{5}{2}$ cm;   (c) $7\frac{1}{2}$ cm;   (d) 8 cm;   (e) 14 cm

8. Let X and Y be points contained in the disk of radius r around the point P.
   Explain why $d(X, Y) \leqq 2r$. Use the Triangle Inequality.

EXPERIMENT 1-2

We have seen in the previous section that the phrase "if and only if" allows us to condense two separate "if-then" statements into a single sentence. For example, consider the statement:

"An integer has a zero in the units place if and only if it is divisible by 10."

The two "if-then" statements which together are equivalent to the above are:

(1) If an integer is divisible by 10, then it has a zero in the units place.
(2) If an integer has a zero in the units place, then it is divisible by 10.

Both of these statements are true, so the original statement is true.

For each of the following statements, write the two "if-then" statements which are equivalent to it:

(a) A number is even if and only if it is divisible by 2.
(b) $6x = 18$ if and only if $x = 3$.
(c) A car is registered in California if and only if it has California license plates.
(d) All the angles of a triangle have equal measure if and only if the triangle is equilateral.
(e) Two distinct lines are parallel if and only if they do not intersect.

Now consider the following statement: "A number is divisible by 4 if and only if it is even." The two "if-then" statements equivalent to it are:

(1) If a number is even, then it is divisible by 4.
(2) If a number is divisible by 4, then it is even.

Statement (2) is true (think about it!), but statement (1) is not (the number 10 is even but not divisible by 4). Therefore the original statement is *not* true, since one half of the "if and only if" condition is false.

Determine whether each of the following are true or false in a similar manner. If false, give an example.

(a) The square of a number is 9 if and only if the number equals 3.
(b) A man lives in California if and only if he lives in the United States.
(c) $a = b$ if and only if $a^2 = b^2$.
(d) Point Q lies in the disk of radius 5 around P if and only if $d(P, Q) \leq 5$.
(e) The integers x, y, z are consecutive if and only if their sum equals $3y$.

In an "if-then" statement, the phrase following the word "if" is often called the **hypothesis** and the phrase following the word "then" is called the **conclusion**. When we interchange the hypothesis and conclusion of an "if-then" statement, we are forming its **converse**. For example, statement (2) above is the converse of statement (1); also, (1) is the converse of (2).

Look back over your work to answer the question: "Is the converse of a true statement always true?"

Give some examples to support your answer. Make up five "if-then" statements of your own and determine whether they are true or not.

1, §3. ANGLES

Consider two rays R_{PQ} and R_{PM} starting from the same point P. These rays separate the plane into two regions, as shown in Figure 1.19:

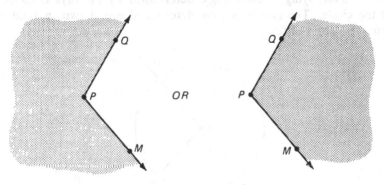

Figure 1.19

Each one of these regions together with the rays will be called an **angle** determined by the rays.

Note. You may already be familiar with the definition of an angle as "the union of two rays having a common vertex". We have chosen a different convention for two reasons. First, people do tend to think of one or the other sides of the rays when they meet two rays as pictured above; they do not think neutrally. Second, and more importantly, when we want to measure angles later, and assign a number to an angle, as when we shall say that an angle has 30 degrees, or 270 degrees, adopting the definition of an angle as the union of two rays would not provide sufficient information for such purposes, and we would need to give additional information to determine the associated measure. Thus it is just as well to incorporate this information in our definition of an angle.

Given two rays as indicated in Figure 1.20, there is a simple notation to distinguish one angle from the other.

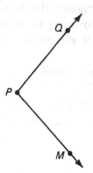

Figure 1.20

We draw a circle whose center is point P, as in Figure 1.21(a). The portion of the circle lying in each angle determined by the rays is called an **arc** of the circle. The two arcs thus determined are shown in bold type in Figure 1.21(b) and (c).

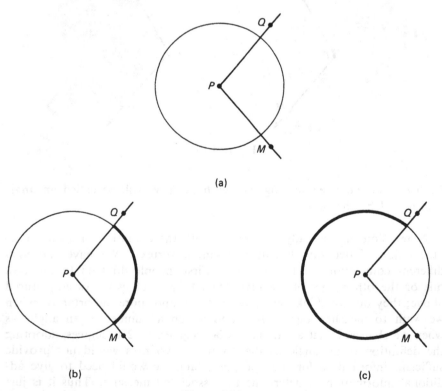

Figure 1.21

Since each arc lies within one of the angles, by drawing one or the other arc, we can indicate which angle we mean, as in Figure 1.22.

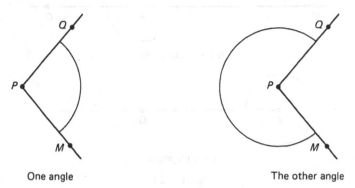

One angle   The other angle

Figure 1.22

We shall use the notation

$$\angle QPM$$

for either one of the two angles determined by the rays R_{PQ} and R_{PM}. The context will always need to be used to determine which one of the two angles is meant. For instance, if we draw the figure as in Figure 1.21(b) or 1.22(a), we mean the angle containing the arc of circle as shown. In another context, when dealing with a triangle as on the Figure 1.23.

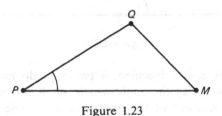

Figure 1.23

by $\angle QPM$ we mean the angle which contains the triangle, also shown with an arc in the figure. We shall also use the abbreviated

$$\angle P$$

instead of $\angle QPM$, if the reference to the rays R_{PQ} and R_{PM} is clear.

Suppose that Q, P, M lie on the same straight line, and that P lies between Q and M. In this case, we say that the indicated angle $\angle QPM$ is a **straight angle** (Figure 1.24(a)). Observe that the other angle determined by the rays R_{PQ} and R_{PM} is also a straight angle.

Suppose $R_{PQ} = R_{PM}$. Then the angle shown in Figure 1.24(b) is called a **full angle**.

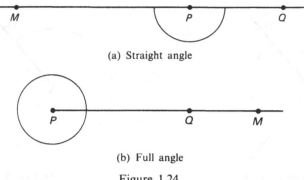

(a) Straight angle

(b) Full angle

Figure 1.24

Just as we used numbers to measure distance, we can now use them to measure angles, provided that we select a unit of measurement first. This can be done in several ways. Here we discuss the most elementary way.

The unit of measurement which we select here is the **degree**, such that the full angle has 360 degrees. We abbreviate "degrees" by using a small circle to the upper right of the number, and so we can write 360° to mean "360 degrees". Since the straight angle divides the full angle into two equal parts, the straight angle has 180°, as shown in Figure 1.25.

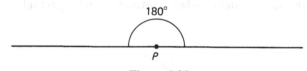

Figure 1.25

Usually, unless otherwise specified, if two rays do not form a straight angle, by $\angle QPM$ we mean the angle which has less than 180°. We shall see later that the measures of the angles of a triangle add up to 180°, and each angle of a triangle has less than 180°.

We define a **right angle** to be an angle whose measure is half the measure of a straight angle; in other words, a right angle is an angle of 90 degrees.

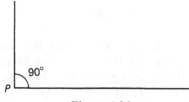

Figure 1.26

With our conventions for abbreviating the notation for angles, this 90° angle would also be denoted by

$$\angle P.$$

An angle that has one degree looks like this:

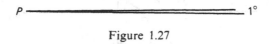

$$P \text{———————————} 1°$$

Figure 1.27

The measure of an angle QPM will be denoted by $m(\angle QPM)$. To say that $\angle QPM$ has 50° means the same thing as

$$m(\angle QPM) = 50°.$$

We define angles $\angle P$ and $\angle R$ to be **supplementary** if

$$m(\angle P) + m(\angle R) = 180°.$$

We have drawn two supplementary angles in Figure 1.28.

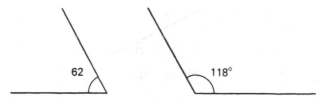

62   118°

Figure 1.28

An important example of supplementary angles is obtained by drawing a line L, a point O on L, and a ray R_{OM} with vertex O, as in Figure 1.29.

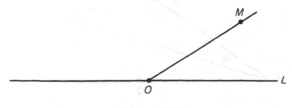

Figure 1.29

The ray separates one side of the staight angle into two angles which are supplementary. If we label these angles $\angle 1$ and $\angle 2$ as in Figure 1.30 we have that $m(\angle 1) + m(\angle 2) = 180°$.

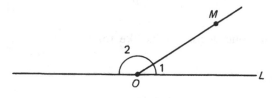

Figure 1.30

We define two angles to be **adjacent** if they have a ray in common. In Figure 1.30, the angles $\angle 1$ and $\angle 2$ are adjacent. In Figure 1.31, $\angle MPR$ and $\angle RPQ$ are adjacent. Clearly,

$$m(\angle MPR) + m(\angle RPQ) = m(\angle MPQ).$$

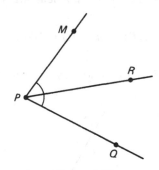

Figure 1.31

We define the **angle bisector** to be the ray which divides the angle into two adjacent angles having the same measure (Figure 1.32).

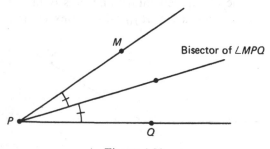

Figure 1.32

Example. An angle of 10° cuts out an arc A on a circle whose circumference is equal to 15 cm. How long is this arc?

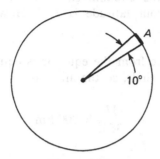

Figure 1.33

We know that the full angle has 360°. Hence the angle of 10° is a fraction of the full angle, namely

$$\frac{10}{360} = \frac{1}{36}.$$

Hence the length of arc cut out is equal to that fraction of the full circumference, that is:

$$\text{length of arc } A = \frac{1}{36} \cdot 15 = \frac{15}{36}.$$

This is a correct answer. You may sometimes wish to simplify the fraction and get the answer $\frac{5}{12}$, which is also correct.

Example. Let us represent the earth by a sphere. Let us draw a great circle through the North Pole N and your home town T, as shown on the Figure 1.34.

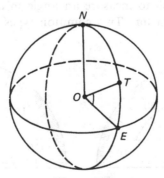

Figure 1.34

Let E be the point on the equator on this great circle. Let O be the center of the earth. The angle $\angle EOT$ is called the **latitude** of your home town. Suppose that the circumference of the great circle has length 40,000 km, and that your latitude is 37°. How far are you from the equator?

Answer. The distance from the equator is equal to the fraction $\frac{37}{360}$ of the length of the great circle, so the answer is

$$\frac{37}{360} \cdot 40,000 \text{ km.}$$

This can be simplified if you wish, to

$$\frac{37}{9} \cdot 1,000 \text{ km.}$$

Both answers are correct.

Remark on Figure 1.34. The equator is drawn as if we were looking at the sphere from above. This being the case, strictly speaking we would then see the North Pole not on the top of the figure, as drawn, but somewhat lower. Nevertheless, we have drawn the figure with the North Pole at the top, despite the inconsistency, because visually we find that the figure gives a better representation of reality. Controversies about such drawings are very classical, and have occurred before, but the practice we follow is adopted frequently, in many countries, for books in astronomy and geometry. It may provide an interesting topic for discussion in class whether the figure is more effective as we have drawn it, or whether it should be drawn otherwise. Look up books on astronomy in a library, and see how they draw similar figures.

Note. When we wish to measure an angle in a picture, we use an instrument called a **protractor**. Two common types are illustrated below in Figure 1.35:

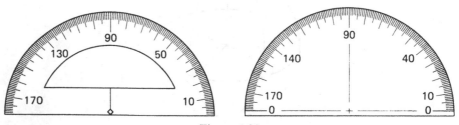

Figure 1.35

To measure a given angle, place the "center" of the protractor, which is usually indicated by a small arrow or cross, on the vertex, and align the 0° mark along one of the rays, as shown in Figure 1.36:

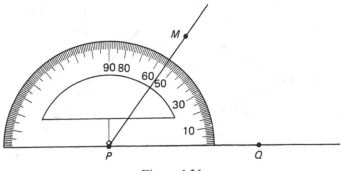

Figure 1.36

Read on the scale where the other ray crosses the protractor. In the example, $m(\angle P) = 55°$. Check that you have read the correct scale on the protractor by estimating the size of the angle—whether it's more or less than 90° and whether it's nearer to 0° or 180°. To measure an angle with more than 180°, measure the other angle determined by the rays and subtract from 360°.

To draw an angle of a given size, draw a ray, place the center of the protractor at the vertex, and make a mark opposite the appropriate number on the scale. Figure 1.37 illustrates how we would draw an angle of 110°.

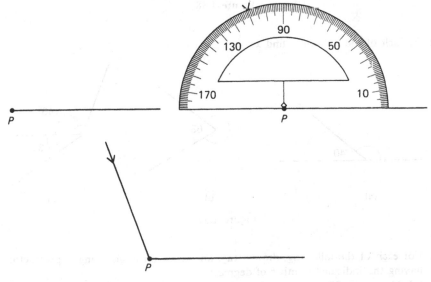

Figure 1.37

1, §3. EXERCISES

1. Name the indicated angles:

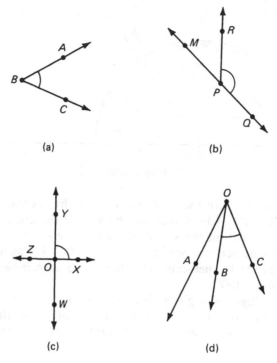

Figure 1.38

2. In each of the following, find x:

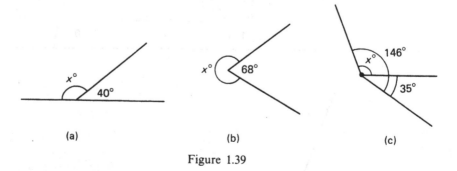

Figure 1.39

3. For each of the following, draw a ray, and draw an angle using a protractor having the indicated number of degrees:
   (a) 45°   (b) 90°   (c) 142°   (d) 192°   (e) 270°

4. Without using a protractor, match the pictures of the angles with their measures:

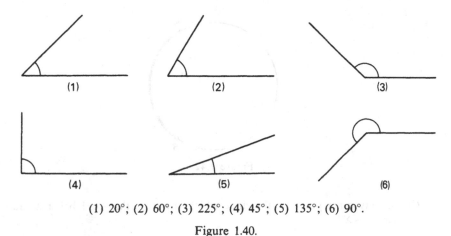

(1) 20°; (2) 60°; (3) 225°; (4) 45°; (5) 135°; (6) 90°.

Figure 1.40.

5. In the Figure 1.41, $\angle P$ determines an arc of the circle centered at P. If the circumference of the circle = 36, how long is the arc if $\angle P = 90°$?

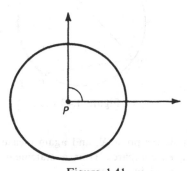

Figure 1.41

6. Same question as above, except how long is the arc if $\angle P$ has:
   (a) 45°   (b) 180°   (c) 60°   (d) $x°$

7. There are 360 degrees of longitude on the earth. How "long" is *one* degree longitude at the *equator* if the circumference of the earth is 40,000 km?

8. Suppose your home town has latitude 43°. Measure distance from your home town along a great circle. How far would you have to travel if you go
   (a) to the equator   (b) to the North Pole?
   [*Hint*: Going to the North Pole, referring to Figure 1.34, what is the measure of the angle $\angle NOT$?]

9. Look up in an atlas the *latitude* of your home city or town (to the nearest degree). Now figure out the distance you would have to travel if you headed directly north to the North Pole.

10. Draw a circle on your paper using a compass; make it a good large circle. Mark off an arc with endpoints *A* and *B* as shown:

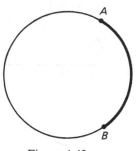

Figure 1.42

Choose any other point on the circle which is not on the arc, label it *X*, and measure ∠*AXB*.

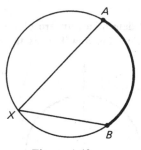

Figure 1.43

Choose other locations for point *X*, and again measure ∠*AXB*. How do the measures of the angles compare? (For a statement and proof of a general theorem, see Theorem 5-3 in Chapter 5.)

11. Using a protractor, find the measures of the three indicated angles.

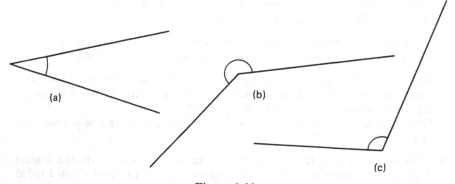

Figure 1.44

12. Draw a ray R_{PQ} horizontally. Using a protractor, draw a second ray R_{PM} such that angle QPM has:
    (a) 60°   (b) 120°   (c) 135°   (d) 160°
    (e) 210°   (f) 225°   (g) 240°   (h) 270°

13. Write the converse of the following statement: "If $m(\angle QPM) = 0°$, then points Q, P, and M lie on the same line." Is the converse TRUE?

CONSTRUCTION 1-2

To draw the angle bisector of a given angle.

You are given angle A, and you wish to construct the ray that bisects it.

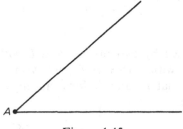

Figure 1.45

Choose a convenient setting of your compass, and with the point on the vertex A, draw an arc crossing the two rays which determine the angle. Label the points where the arc intersects the rays P and Q.

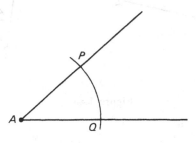

Figure 1.46

Place the tip of the compass at P, and draw arc g, as below. Still using the same compass setting, place the tip at Q and draw arc h. Label the point where arcs g and h intersect point M. The ray starting at A

through M is the angle bisector. We shall see *why* this works later in the course.

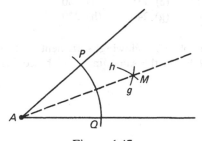

Figure 1.47

CONSTRUCTION 1-3

Duplicating an angle

Given an angle formed by two rays K and L with the same vertex at O, and given a ray K' with vertex at O', we wish to construct a ray L' with vertex at O' such that K' and L' form an angle with the same measure as the given angle.

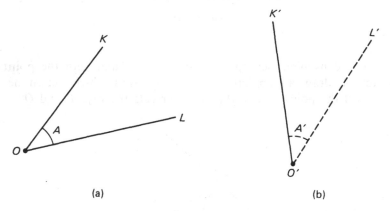

(a)   (b)

Figure 1.48

In the figure, angle A is given, K' is given, and we want to construct L'. Select a point $P \neq O$ on ray K. Set the distance between the tips of your compass to be $d(O, P)$. Put the tip of your compass on O and draw an arc through P, intersecting line L in a point which you label Q as shown on the figure.

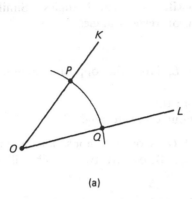
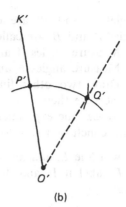

(a)   (b)

Figure 1.49

Without changing the setting of the compass, place the tip at O' and draw an arc intersecting ray K' in a point which you label P'.

Next set the compass tips at a distance equal to $d(P, Q)$. With the tip placed on point P', draw an arc intersecting the previous arc in a point Q'. *Then the angle $\angle P'O'Q'$ has the same measure as $\angle POQ$.* Observe that the triangles $\triangle POQ$ and $\triangle P'O'Q'$ are isosceles. In fact, our construction shows that:

$$|PO| = |P'O'| = |QO| = |Q'O'|$$

writing $|PO|$ for $d(P, O)$, and similarly for other distances. Furthermore,

$$|PQ| = |P'Q'|.$$

Thus the two triangles $\triangle POQ$ and $\triangle P'O'Q'$ have corresponding sides having the same length. It is usually taken as a postulate that the corresponding angles of such triangles have the same measure. This matter is discussed at length in Chapter 7.

EXPERIMENT 1-3

1. Draw two lines K and L which intersect. These lines give rise to pairs of angles A, A' and B, B' as shown:

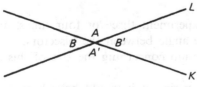

Figure 1.50

Angles A and A' are called **opposite**, or **vertical** angles. Similarly, angles B and B' are called **opposite** or **vertical** angles.

(a) Measure angles A and A'.

(b) Measure angles B and B'.

(c) Draw two other lines K and L, label the opposite angles, and measure them.

(d) Repeat the experiment one more time.

What conclusion can you reach about opposite angles?

2. Draw a line L, and locate points Q, O, R on it. Choose any point not on L, label it P, and draw ray R_{OP}. Below are two possible illustrations:

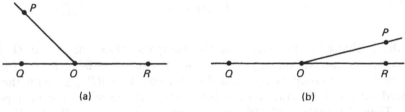

Figure 1.51

You have created two, adjacent angles, $\angle QOP$ and $\angle ROP$. Using your ruler and compass, bisect each of these angles, and measure the angle determined by the bisectors, as indicated:

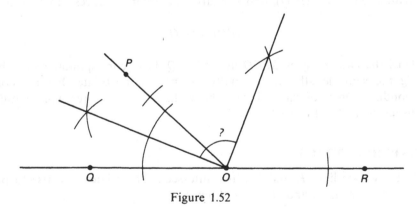

Figure 1.52

(a) Repeat this experiment three or four times, and keep a record of the size of the angle between the bisectors.

(b) State a conclusion concerning the size of this angle.

In the previous experiment it would have been possible to reach conclusions *without* appealing to constructions or measurements. Instead, we could have presented an "argument" that these conclusions must be true, based on what we have learned so far in geometry and what we know of algebra. You should now try to formulate such arguments which demonstrate that your conclusions are true. Here are some hints to get you started:

Exercise 1. What does $m(\angle A) + m(\angle B) = ?$ Why?
What does $m(\angle A') + m(\angle B) = ?$ Why?

From these two equations and some algebra, what can you conclude about $m(\angle A')$ and $m(\angle A)$?

Exercise 2. Look at the two, original adjacent angles.
What does $m(\angle QOP) + m(\angle ROP) = ?$ Why?

Let $m(\angle QOP) = x°$. What is the measure of $\angle ROP$ in terms of x? The angle between the bisectors is the sum of $\frac{1}{2}$ of $\angle ROP$ plus $\frac{1}{2}$ of $\angle QOR$. Express this sum algebraically and simplify.

Look over your answers to the questions above and write out a complete argument in paragraph form "proving" your conclusion for each of the two exercises, just as a lawyer presents evidence to support his case in court. Use full English sentences and correct mathematical statements. In the next section, you will find a statement and proof, but don't look them up without first trying it out for yourself.

1, §4. PROOFS

In Experiment 1-3 you made a first try at writing a "proof". What do we mean by a "proof" in general? More specifically, what do we mean by a "proof of a statement"? Let S denote the statement. In Experiment 1-3 you went through several steps. At each step, you made an assertion, which was either an assumption, or followed from previous statements by rules of reasoning. A proof in general is nothing more than that.

A **proof** of a statement S is a sequence of assertions such that each assertion is an assumption, or is a consequence of previous assertions by the rules of reasoning, and statement S is the last assertion of the sequence.

A lot of the assumptions which we make are due to previous knowledge which you acquired in other courses, for instance, the basic rules of algebra, like $a + b = b + a$, or substituting equals for equals in an equation.

Of course, a proof relies on accepted results. In each case, one should try to make explicit what these results are, what assumptions are made, and what references are used to close up a logical chain of reasoning. Readers are entitled to ask for a justification of all assertions, and reasons for the deductions. These reasons can be by logical reasoning or specific references to past results.

A **theorem** is a statement which has been proved on the basis of established knowledge. If you were successful in Part 1 of Experiment 1-3, you proved the following statement.

Theorem 1-1. *Opposite angles have the same measure.*

A typical write up of the proof might look like this:

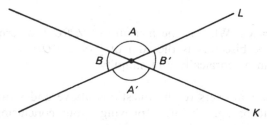

Figure 1.53

Proof. Let lines K, L intersect at a point P, forming angles $\angle A$, $\angle A'$, $\angle B$, $\angle B'$. Since $\angle A$ and $\angle B$ together form a straight angle, we have

$$m(\angle A) + m(\angle B) = 180°.$$

Since $\angle B$ and $\angle A'$ form a straight angle, we have

$$m(\angle B) + m(\angle A') = 180°.$$

Combining these statements, we see that

$$m(\angle A) + m(\angle B) = m(\angle B) + m(\angle A').$$

Subtracting $m(\angle B)$ from both sides, we obtain

$$m(\angle A) = m(\angle A').$$

This proves the theorem.

Though you did some measuring in Experiments 1-3, measurement with a protractor or a ruler is not valid for a *proof* for the following reasons:

1. The measurement is not totally accurate. It is only an approximation. Of course this approximation is good enough for some practical applications, but we are interested here also in exact statements. Most protractors and rulers would anyhow be good only up to two or three decimal points of accuracy.
2. Measurement is only good for specific cases. Suppose you have determined that a statement is true for angles with 54°. What about angles which have 55°, or 65°, etc.? You would have to do an infinite amount of measuring to check other angles, and that you cannot do.

Though we often rely on pictures and measurements to give us a clue about the way to prove some statement, nevertheless a proof will only involve logical reasoning.

In the proof of Theorem 1-1, reasons are given for nearly every statement made during the proof. Notice the use of the word "since". For the last statement, we point out that we are "subtracting". Statements which are obvious from your knowledge of algebra or arithmetic do not require explanation. An example in the above proof occurs when we state that

$$m(\angle A) + m(\angle B) = m(\angle B) + m(\angle A').$$

This is obvious once you know that each side of the equation is equal to 180°, as stated at the beginning of the proof.

One final remark about proof. You might well ask why we don't prove *everything*. Though we proved that "opposite angles are equal", we just assumed as fact such notions as **PAR 1**, **PAR 2**, the Triangle Inequality, and other statements earlier in the book. Why don't we prove these? On the other hand, why did we not just assume "opposite angles are equal", label it **ANG 1**, and be done with it?

The reason that we don't prove everything is that we can't; we have to start with something. There must be some statements which we accept as true without proof, upon which we can build further knowledge. Such statements are called **postulates** or **axioms**.

We *prove* theorems, rather than just call everything we feel is true a postulate, because our brains would not feel comfortable accepting everything on faith. Perhaps you would be willing to accept Theorem 1-1 on faith, but there are many results in geometry that are very surprising, and you will find yourself saying "prove it!". Statements like **PAR 1** and **PAR 2** are not surprising, and so we accept them as facts upon which to build. Of course, much of this is a matter of taste. A statement that is ridiculously obvious to one person might arouse doubts in someone else.

We hope that in this book we have found a good balance between what we assume as postulates, and what we prove as theorems.

Before you try your hand at writing some more proofs, we give other examples.

Example. In the figure,

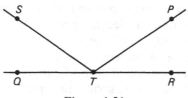

Figure 1.54

$m(\angle STQ) = m(\angle PTR)$. Prove that $m(\angle PTQ) = m(\angle STR)$.

Proof. We know that $m(\angle STQ) = m(\angle PTR)$. Adding $m(\angle STP)$ to both sides of the equation, we have:

$$m(\angle STQ) + m(\angle STP) = m(\angle PTR) + m(\angle STP).$$

Since

$$m(\angle STQ) + m(\angle STP) = m(\angle PTQ)$$

and

$$m(\angle PTR) + m(\angle STP) = m(\angle STR),$$

substitute in the previous equation to get:

$$m(\angle PTQ) = m(\angle STR).$$

This proves what we wanted.

Example. In the figure,

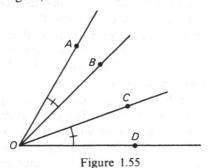

Figure 1.55

assume that

$$m(\angle AOB) = m(\angle COD).$$

Prove that $m(\angle AOC) = m(\angle BOD)$.

Proof. Angles $\angle AOC$ and $\angle COD$ together form $\angle AOD$. Hence

$$m(\angle AOC) + m(\angle COD) = m(\angle AOD).$$

Similarly,

$$m(\angle AOB) + m(\angle BOD) = m(\angle AOD).$$

The right-hand sides of these equations are equal. Therefore, so are the left-hand sides, that is,

$$m(\angle AOC) + m(\angle COD) = m(\angle AOB) + m(\angle BOD).$$

By assumption, we can cancel $m(\angle COD)$ and $m(\angle AOB)$ which are equal. We are then left with

$$m(\angle AOC) = m(\angle BOD)$$

as was to be proved.

As you write proofs, remember two things:

1. Any postulate or any theorem that has already been proved is acceptable as known facts.
2. As you write, imagine that someone else will read your work, and will want to follow your argument. This will help put your proof in good form.

1, §4. EXERCISES

1. What is a postulate? Give an example of
   (a) a postulate used in geometry;
   (b) a postulate used in algebra.

2. Three lines intersect at point O as shown in Figure 1.56. Angle QOR has $55°$ and angle POS has $90°$.

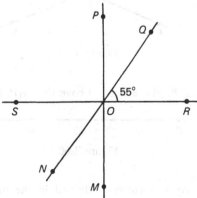

Figure 1.56

   (a) What is $m(\angle NOM)$?
   (b) Name another angle having the same measure as $\angle SOQ$.

3. Two lines intersect at a point P as shown in Figure 1.57. If $m(\angle WPX)$ is twice the measure of $\angle WPY$, find the number of degrees in $\angle YPV$.

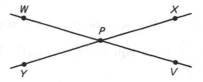

Figure 1.57

4. In Figure 1.58, R_{OB} is the bisector of $\angle AOC$ and R_{OC} is the bisector of $\angle BOD$. Prove that $m(\angle AOB) = m(\angle COD)$.

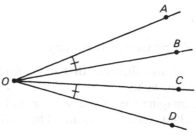

Figure 1.58

5. Three lines intersect at a point O as indicated in Figure 1.59. Line QT bisects $\angle POR$. Show that it bisects $\angle SOV$ as well.

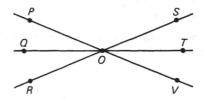

Figure 1.59

6. In the Figure 1.60, $d(P, Q) = d(R, S)$. Prove that $d(P, R) = d(Q, S)$.

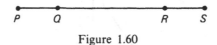

Figure 1.60

7. Let X and Y be any two points contained in the disc of radius r around point P. Using the triangle inequality, prove that

$$d(X, Y) \leqq 2r.$$

8. Line L intersects lines K and U as shown in Figure 1.61. If $\angle 1$ is supplementary to $\angle 2$, prove that $m(\angle 3) = m(\angle 4)$.

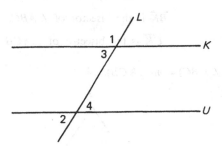

Figure 1.61

9. In triangle ABC shown in Figure 1.62, $m(\angle CAB) = m(\angle CBA)$. In triangle ABD, $m(\angle DAB) = m(\angle DBA)$. Prove that

$$m(\angle CAD) = m(\angle CBD).$$

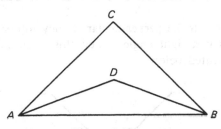

Figure 1.62

10. The lines L_{AB} and L_{BC} are cut by the line L_{PS} at points Q and R. Assume $m(\angle b) = m(\angle c)$. Prove that $m(\angle a) = m(\angle d)$.

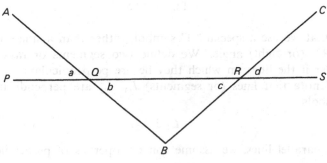

Figure 1.63

11. Points P, B, C, and Q lie on a line as shown in Figure 1.64. Furthermore:

$$m(\angle x) = m(\angle y),$$

$\overline{BK}$ is the bisector of $\angle ABC$,

$\overline{CK}$ is the bisector of $\angle ACB$.

Prove that $m(\angle KBC) = m(\angle KCB)$.

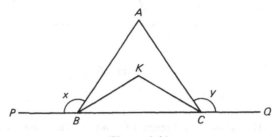

Figure 1.64

1, §5. RIGHT ANGLES AND PERPENDICULARITY

We define two lines to be **perpendicular** if they intersect, and if the angle between the lines is a right angle. Then this angle has 90°. Perpendicular lines are illustrated below.

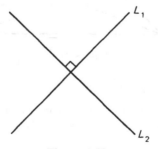

Figure 1.65

Notice that we use a special "⌐" symbol, rather than a small arc, to indicate a 90° (or right) angle. We define two segments or rays to be **perpendicular** if the lines on which they lie are perpendicular.

To denote that lines (or segments) L_1, L_2 are perpendicular, we use the symbols

$$L_1 \perp L_2.$$

As with parallel lines, we assume some properties of perpendicular lines, namely:

PERP 1. *Given a line L and a point P, there is one and only one line through P, perpendicular to L.*

This is illustrated in Figure 1.66. Can you state the corresponding postulate for parallel lines?

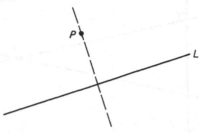

Figure 1.66

PERP 2. *Given two parallel lines L_1 and L_2. If a line K is perpendicular to L_1, then it is perpendicular to L_2.*

PERP 2 is illustrated in Figure 1.67.

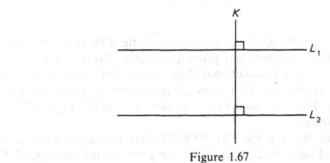

Figure 1.67

Using these two new postulates, we can prove a theorem relating perpendiculars and parallels.

Theorem 1-2. *If L_1 is perpendicular to line K, and L_2 is also perpendicular to line K, then L_1 is parallel to L_2.*

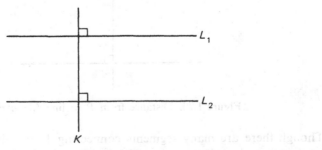

Figure 1.68

Proof. Either L_1 and L_2 are parallel or they're not. Suppose they are *not* parallel. Then by **PAR 1** we know they intersect in one point, say P, and we have a picture as below:

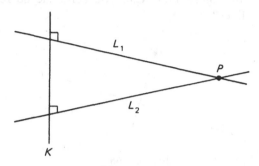

Figure 1.69

But here we have two lines passing through point P, each perpendicular to line K (remember it is *given* that they are perpendicular to K). This situation contradicts **PERP 1**, a postulate. Therefore lines L_1 and L_2 must be parallel.

Incidentally, the proof for Theorem 1-2 is a little different from those we have done before. Rather than proving directly that L_1 and L_2 are parallel, we suppose for a moment that they are *not*, and then show that this leads to a contradiction. This is an effective form of argument, used often in everyday life, and sometimes known as "indirect proof" or **"proof by contradiction"**.

Let P be a point and L a line. By **PERP 1** there is exactly one line K through P, perpendicular to L. Let M be the point of intersection of K and L. We define the **distance from P to L** to be the length of $\overline{PM}$. In other words, it is the length of the perpendicular segment from P to L.

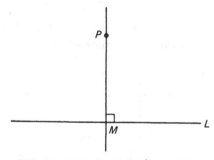

Figure 1.70. Distance from P to line $L = d(P, M)$.

Though there are many segments connecting P and line L, **PERP 1** tells us that there is only one *perpendicular*, and thus there is no confusion.

The following postulate formalizes the intuitive notion that the "distance" between two parallel lines is constant "all the way along the lines".

PD (Parallel Distance). *Let L_1 and L_2 be two parallel lines, and let P and Q be any two points on L_1. The distance from P to L_2 is equal to the distance from Q to L_2.*

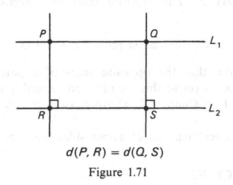

$$d(P, R) = d(Q, S)$$

Figure 1.71

Some final definitions:

Let P, Q, M, N be four points which determine the four-sided figure consisting of the four sides $\overline{PQ}$, $\overline{QM}$, $\overline{MN}$, and $\overline{NP}$. Any four-sided figure in the plane is called a **quadrilateral**. If the opposite sides of the quadrilateral are parallel, that is, if $\overline{PQ}$ is parallel to $\overline{MN}$ and $\overline{PN}$ is parallel to $\overline{QM}$, then the figure is called a **parallelogram**:

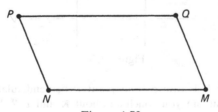

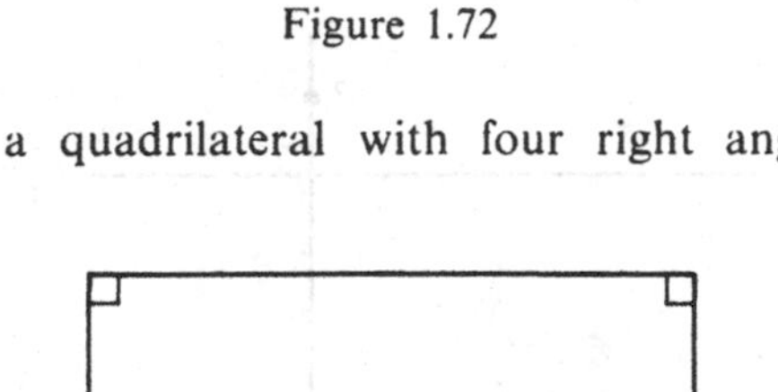

Figure 1.72

A **rectangle** is a quadrilateral with four right angles, as shown in Figure 1.73.

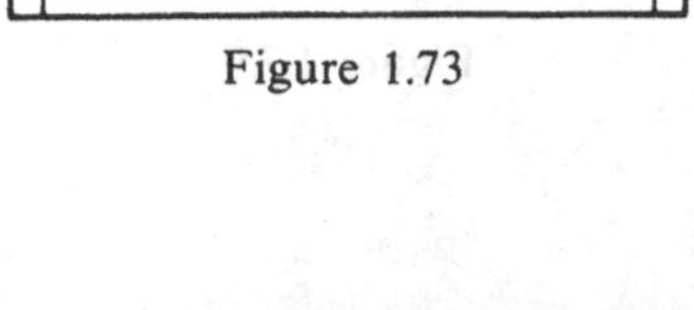

Figure 1.73

In the definition of a rectangle we only assumed something about the angles. The following basic properties will be used constantly.

Basic Property 1. *The opposite sides of a rectangle are parallel, and therefore a rectangle is also a parallelogram.*

This is a direct consequence of Theorem 1-2.

Basic Property 2. *The opposite sides of a rectangle have the same length.*

This follows from the first basic property and **PD**.

It is also true that the opposite sides of a parallelogram have the same length, but to prove this we need additional axioms. You can look up Theorem 7-1 in Chapter 7, §2 right away if you want to see how a proof goes.

A **square** is a rectangle all of whose sides have the same length.

1, §5. EXERCISES

1. In Figure 1.74 line K is perpendicular to line V, and line L is perpendicular to line V. What can you conclude about lines K and L? Why?

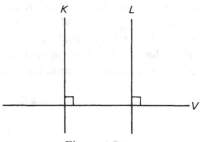

Figure 1.74

2. In Figure 1.75 line K is drawn from point P perpendicular to L_1. If L_1 and L_2 are parallel, what can you conclude about K and L_2? Why?

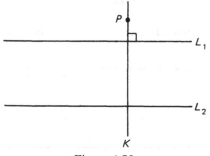

Figure 1.75

3. In Figure 1.76,

$$\overline{PR} \text{ is perpendicular to } \overline{PT};$$
$$\overline{PQ} \text{ is perpendicular to } \overline{PS}.$$

Prove that $m(\angle a) = m(\angle b)$.

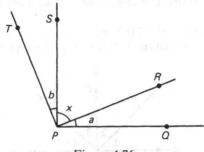

Figure 1.76

4. Below is an example of a non-mathematical proof by contradiction:

Johnny wants to go to the store after dinner. His mother says no, he should stay home, do his work, and maintain his straight-A average. Johnny argues: "Suppose I don't go to the store tonight. Then I won't be able to get a protractor. Tomorrow, I'll be without one in geometry class, and the teacher will get mad. As a result, I'll get an F. Since that is an intolerable thing, you must let me go to the store."

Give another example of a "proof by contradiction" that you might have used some time in your life.

5. Below is an alternate definition for a rectangle:

"A rectangle is a parallelogram with one right angle."

The figure below illustrates this definition; $ABCD$ is a parallelogram where A is the right angle.

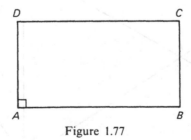

Figure 1.77

Use **PERP 2** to show that all the angles are right angles.

Thus you see that this "new" definition is really saying the same thing as our original one.

6. Refer to Figure 1.78 to answer the following. We assume $L_{CD} \perp L_{AB}$.
   (a) $m(\angle 1) + m(\angle 2) = \underline{\hspace{1.5cm}}°$.
   (b) If $m(\angle 3) = 50°$, then $m(\angle 4) = \underline{\hspace{1.5cm}}°$.
   (c) Is $\angle AOT$ the supplement of $\angle TOB$? $\underline{\hspace{1.5cm}}$.
   (d) $m(\angle 1) + m(\angle 2) + m(\angle 3) + m(\angle 4) = \underline{\hspace{1.5cm}}°$.
   (e) If $m(\angle 4) = 23°$, then $m(\angle 3) = \underline{\hspace{1.5cm}}°$.
   (f) Name, using numbers, two angles that are adjacent to $\angle 2$. $\underline{\hspace{1.5cm}}$ and

   $\underline{\hspace{1.5cm}}$.

   (g) If $m(\angle 1) = 32°$, then $m(\angle TOB) = \underline{\hspace{1.5cm}}°$.
   (h) Must R_{OT} be perpendicular to R_{OS} if $m(\angle 1) + m(\angle 4) = 90°$? $\underline{\hspace{1.5cm}}$.

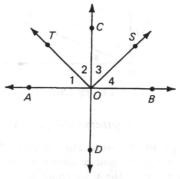

Figure 1.78

7. Suppose L_1 and L_2 are both perpendicular to L_3, and they both intersect L_3 at point P. What can you conclude about L_1 and L_2?

8. In quadrilateral $PBQC$, assume that $\angle PBQ$ and $\angle PCQ$ are right angles, and that $m(\angle x) = m(\angle y)$. Prove that

$$m(\angle ABQ) = m(\angle DCQ).$$

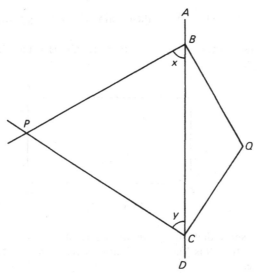

Figure 1.79

9. The axiom **PAR 3** can actually be proved from axioms **PAR 1** and **PAR 2** using a proof by contradiction.

  Given L_1 parallel to L_2, and L_2 parallel to L_3. Suppose that L_1 and L_3 are *not* parallel. Use **PAR 1** and **PAR 2** to derive a contradiction.

  Once you have done this proof, **PAR 3** can be considered a theorem, since we have proved it from earlier facts.

CONSTRUCTION 1-4

To construct the perpendicular from a point on a given line.

Given line L, you wish to construct a line perpendicular to L through a point P on L.

  A perpendicular forms a 90° angle, which is $\frac{1}{2}$ of a straight angle, and so the construction essentially amounts to bisecting a straight angle.

  Choose a convenient radius, place the tip of the compass on P, and draw an arc, labeling points M and Q as shown in Figure 1.80.

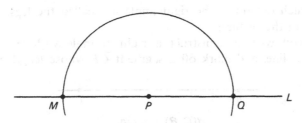

Figure 1.80

Using a larger compass radius, place the tip at Q and draw an arc above P, and then place the tip at M and draw another arc intersecting the first. The line through P and this intersection is the perpendicular.

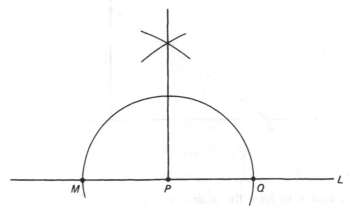

Figure 1.81

CONSTRUCTION 1-5

To construct a right triangle given the lengths of the two legs.

A **right triangle** is a triangle with a 90° (right) angle, as illustrated:

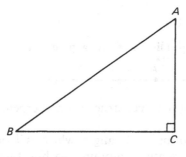

Figure 1.82

The sides which determine the right angle are called the **legs**. These are $\overline{CA}$ and $\overline{CB}$ in the picture.

Suppose you wish to construct a right triangle with legs 6 cm and 8 cm. Draw a line, and mark off a segment $\overline{CB}$ whose length is 6 cm.

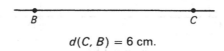

$$d(C, B) = 6 \text{ cm.}$$

Construct the line through C perpendicular to the first line, and locate point A such that $d(C, A) = 8$ cm.

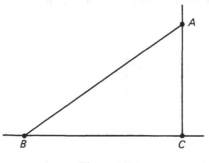

Figure 1.83

Connect A and B to form the desired triangle.

EXPERIMENT 1-4

1. Construct a right triangle with legs 8 cm and 11 cm. Label it ABC, where C is the right angle. Use your protractor to measure angle A. Measure angle B.

$$m(\angle A) + m(\angle B) = ?.$$

2. Repeat Exercise 1 again, only this time make the legs of the right triangle 10 cm and 13 cm. (If you see a method of getting a right triangle using the squared-off corner of a piece of paper, use it!) Repeat this exercise once more, using legs 2 cm and 9 cm. What can you conclude about $m(\angle A) + m(\angle B)$?

3. Cut out the first triangle you made is Exercise 1. Construct another right triangle with legs 8 cm and 11 cm and cut it out. Can you make this other triangle a different shape but still using the legs 8 cm and 11 cm?

4. Take two of the triangles you were working with in Exercise 3, and put them together as shown below in Figure 1.84.

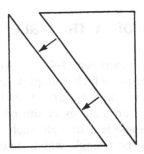

Figure 1.84

What figure will be formed when they are joined?

5. Using a straight edge, draw any arbitrary triangle, and label the angles A, B, and C, as shown below:

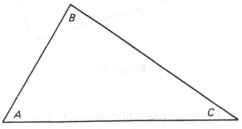

Figure 1.85

Cut it out, tear off angles *A* and *C*, and place them alongside angle *B* as shown:

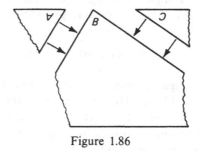

Figure 1.86

Repeat this with three or four other arbitrary triangles. What do you notice?

6. Explain how constructing a perpendicular from a point on a line resembles bisecting an angle.

1, §6. THE ANGLES OF A TRIANGLE

In Construction 1-5 and Experiment 1-4, you worked with a special kind of triangle, the right triangle, which will play an important role in what follows. Many properties of arbitrary triangles, or other geometric figures, can be reduced to an analysis of the properties of right triangles.

Recall that a **right triangle** is a triangle such that one of its angles is a right angle. The sides of the triangle which determine the right angle are called the **legs**; the side other than the legs is called the **hypotenuse**. In Figure 1.87 below, $\overline{PQ}$ and $\overline{PM}$ are the legs, while $\overline{QM}$ is the hypotenuse.

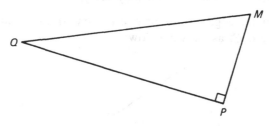

Figure 1.87

In Experiment 1-4, you probably realized that if two right triangles have the same size legs, the two triangles are exact copies of each other.

(If you didn't realize this, go back and cut up some more triangles!) We formalize this notion in the next postulate.

RT (Right Triangle). *Let $\triangle ABC$ and $\triangle XYZ$ be right triangles whose corresponding legs have the same length, that is:*

$$|BC| = |YZ|$$

and

$$|AC| = |XZ|.$$

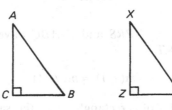

Figure 1.88

Then the hypotenuses have the same length, the corresponding angles are the same size, and the triangles have the same area. In particular,

$$|AB| = |XY|, \quad m(\angle A) = m(\angle X), \quad m(\angle B) = m(\angle Y).$$

We can now prove a property that you most likely noticed in Parts 1 and 4 of Experiment 1-4:

Theorem 1-3. *If A, B are the angles of a right triangle other than the right angle, then*

$$m(\angle A) + m(\angle B) = 90°.$$

Figure 1.89

Proof. Let $PQRS$ be a rectangle where $|PS| = |AC|$ and $|RS| = |BC|$, and draw segment $\overline{PR}$:

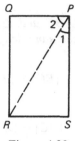

Figure 1.90

Note that right triangles $\triangle PRS$ and $\triangle ABC$ have corresponding legs of the same length. By **RT**,

$$m(\angle 1) = m(\angle A).$$

Since the opposite sides of a rectangle have the same length, we have

$$|QP| = |RS| \qquad \text{and} \qquad |QR| = |PS|.$$

By **RT** then, we have

$$m(\angle 2) = m(\angle B).$$

Since $\angle 1$ and $\angle 2$ form the corner of a rectangle, we have

$$m(\angle 1) + m(\angle 2) = 90°.$$

Substituting, we then obtain:

$$m(\angle A) + m(\angle B) = 90°,$$

thereby proving the theorem.

Example. Let A, B, C be the angles of a right triangle, and suppose that C is the right angle. If $m(\angle A) = 31°$, then we can find $m(\angle B)$, namely

$$m(\angle B) = 90 - m(\angle A) = 90 - 31 = 59.$$

Since our measures of angles are in degrees, we should actually write $m(\angle B) = 59°$.

Remark. If $PQRS$ is a parallelogram, then the segments $\overline{PR}$ and $\overline{QS}$ are called the **diagonals**. For instance, in Figure 1.90 we have used the diagonal of a rectangle. In the case of a rectangle or more generally a parallelogram, a diagonal separates the parallelogram into two triangles. We shall often encounter this way of looking at triangles. In the same way one defines the diagonals of a quadrilateral. For the use of diagonals similar to that of Theorem 1-3, see the proofs of Theorems 3-1 and 3-3 in Chapter 3. You can read those proofs now without going through Chapter 2.

Exercise. Prove that the diagonals of a rectangle have the same length. Is this true for the diagonals of a parallelogram?

In a right triangle, if we know one angle besides the right angle, then we can find the measure of the third angle by a subtraction as in the example.

An interesting situation occurs when two parallel lines are crossed by a third line. This is illustrated in Figure 1.91, where we have also labeled the eight angles formed.

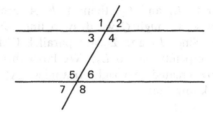

Figure 1.91

Certain pairs of these angles have special names. Angles 2 and 6 are called **parallel angles**. Other instances of parallel angles are angles 4 and 8, angles 1 and 5, and angles 3 and 7.

Angles 4 and 5 are called **alternate angles**, as are angles 3 and 6.

Theorem 1-3 provides the means to prove the following useful theorem.

Theorem 1-4. *Given line K intersecting parallel lines L_1 and L_2. Then parallel angles A and B have the same measure, and so do alternate angles B and A'.*

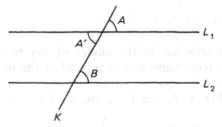

Figure 1.92

Proof. Let Q be a point on K lying above the line L_1 as shown in Figure 1.93.

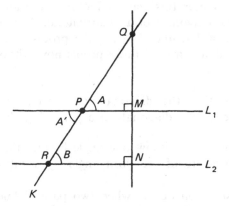

Figure 1.93

If K is perpendicular to L_1 and L_2, then A, B, A' are right angles and we are done. Otherwise, through Q we draw a line perpendicular to L_1, intersecting L_1 at M. Since L_1 and L_2 are parallel, **PERP 2** tells us that this new line is also perpendicular to L_2. We label this point of intersection N. We have now created two right triangles, $\triangle QPM$ and $\triangle QRN$. By Theorem 1-3, we know that

$$m(\angle B) + m(\angle Q) = 90° \qquad \text{(applied to } \triangle QRN)$$

and

$$m(\angle A) + m(\angle Q) = 90° \qquad \text{(applied to } \triangle QPM).$$

Therefore $m(\angle A) = m(\angle B)$. Since opposite angles have the same measure, we have

$$m(\angle A) = m(\angle A')$$

and thus

$$m(\angle B) = m(\angle A').$$

This proves our theorem.

The sum of the measures of the angles of any triangle is a constant, namely 180°. This remarkable fact is proved in the next theorem.

Theorem 1-5. *Let A, B, and C be the angles of an arbitrary triangle. Then*

$$m(\angle A) + m(\angle B) + m(\angle C) = 180°.$$

Proof. Let K be the line through B parallel to $\overline{AC}$, and label the newly formed angles $\angle 1$ and $\angle 2$.

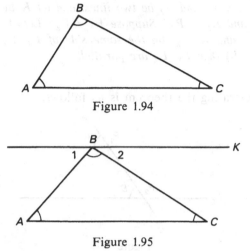

Figure 1.94

Figure 1.95

Since K is parallel to L_{AC}, angles A and 1 are alternate angles, as are angles C and 2. Theorem 1-4 tells us that

$$m(\angle A) = m(\angle 1)$$

and

$$m(\angle C) = m(\angle 2).$$

However, looking at the angles at B, we have

$$m(\angle 1) + m(\angle B) + m(\angle 2) = 180°.$$

Substituting, we finally obtain:

$$m(\angle A) + m(\angle B) + m(\angle C) = 180°.$$

We can now prove the converse of Theorem 1-4. This tells us that in the situation illustrated below, if $m(\angle A) = m(\angle B)$, then L_1 is parallel to L_2.

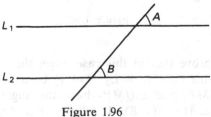

Figure 1.96

This is a very useful test to determine whether two lines are indeed parallel.

Theorem 1-6. *Let L_1 and L_2 be two lines, and let K be a line intersecting L_1 at P_1 and L_2 at P_2. Suppose $P_1 \neq P_2$. Let A, B be the angles between K, L_1 and K, L_2, on the same side of L_1, L_2 respectively. If $m(\angle A) = m(\angle B)$ then L_1, L_2 are parallel.*

The figure illustrating the theorem is as follows.

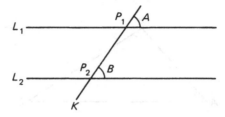

Figure 1.97

Proof. If L_1, L_2 do not intersect, they are parallel and we are done. On the other hand, suppose that L_1, L_2 intersect in a point M, so that possibly the situation looks like this.

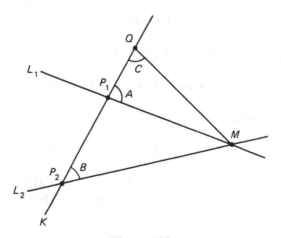

Figure 1.98

We are going to prove that in this case when the lines L_1, L_2 intersect, then in fact we must have $L_1 = L_2$. Let Q be a point on K as shown. The triangles $\triangle QMP_1$ and $\triangle QMP_2$ have the angle C in common, and by hypothesis, $m(\angle A) = m(\angle B)$. It follows that $\angle QMP_1$ and $\angle QMP_2$

have the same measure, because

$$m(\angle C) + m(\angle A) + m(\angle QMP_1) = 180,$$

$$m(\angle C) + m(\angle B) + m(\angle QMP_2) = 180.$$

Therefore P_1 and P_2 lie on the same line. Since we assumed $P_1 \neq P_2$ it follows that the lines L_1, L_2 have two distinct points in common, so $L_1 = L_2$ and the lines are parallel, thus proving our theorem.

Example. We wish to find the angle x in the diagram. The horizontal lines are meant to be parallel.

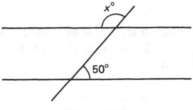

Figure 1.99

There is a supplementary angle for the angle of $x°$ which we label A to obtain the following picture. Then

$$x + m(\angle A) = 180°.$$

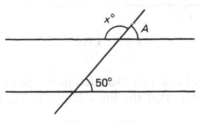

Figure 1.100

By Theorem 1-4 we know that $m(\angle A) = 50°$. Hence we find

$$x + 50 = 180.$$

Subtracting 50 from both sides yields $x = 130$.

Remark. In Construction 1-3 we gave a procedure for duplicating an angle. This can now be applied to the construction of a line parallel to a given line, through a given point P. We are given a line L and a point P, and we wish to construct a line through P, parallel to L.

Draw any line K through P intersecting L in a point which you label O.

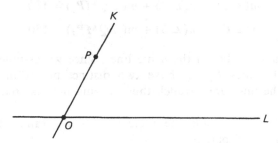

Figure 1.101

Let $\angle a$ be the angle which line K makes with L. The idea of the construction is to draw a line L' through P, such that K makes an angle $\angle b$ with L', and such that

$$m(\angle a) = m(\angle b).$$

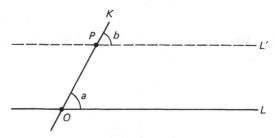

Figure 1.102

This is merely a special case of Construction 1-3 to duplicate the angle a. Then Theorem 1-6 shows that the lines L and L' are parallel.

1, §6. EXERCISES

1. Let A, B, and C be the angles of a right triangle, with A the right angle. Find $m(\angle C)$ when $m(\angle B)$ has the following values:
   (a) $m(\angle B) = 34°$ (b) $m(\angle B) = 60°$ (c) $m(\angle B) = 30°$
   (d) $m(\angle B) = 45°$

2. Let A, B, and C be the angles of any triangle. For each of the following values of $m(A)$ and $m(B)$, find $m(C)$:
   (a) $m(\angle A) = 47°, m(\angle B) = 110°$ (b) $m(\angle A) = 120°, m(\angle B) = 30°$
   (c) $m(\angle A) = 60°, m(\angle B) = 70°$ (d) $m(\angle A) = 90°, m(\angle B) = 54°$

3. Find x in each of the following (if two lines *look* parallel, they're meant to be in this exercise):

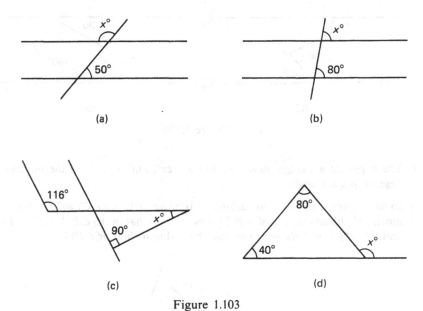

(a)

(b)

(c)

(d)

Figure 1.103

4. Without a protractor, find the value of x in each of the following:

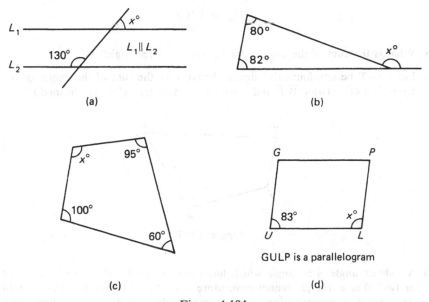

(a)

(b)

GULP is a parallelogram

(c)

(d)

Figure 1.104

5. In each figure below, L_1 is parallel to L_2. Find x. [*Hint*: Draw a line through P parallel to L_1.]

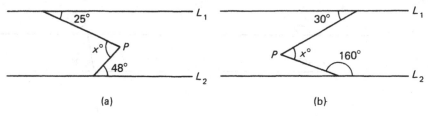

(a)   (b)

Figure 1.105

6. The angles of a triangle have measure $x°$, $2x°$, and $3x°$. Find the number of degrees in each angle.

7. In the figure, $\triangle ABC$ is an arbitrary triangle with $m(\angle C) = 70°$. Furthermore, $\overline{AP}$ bisects $\angle A$ and $\overline{PB}$ bisects $\angle B$. What is $m(\angle P)$? [*Hint*: First find $m(\angle A) + m(\angle B)$, then use half this value to find $m(\angle P)$.]

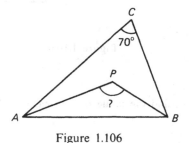

Figure 1.106

8. What is the sum of the angles of a square? of a rectangle?

9. Let $SWAT$ be any four-sided figure. Prove that the sum of the angles of the figure is 360°. (Draw $\overline{WT}$ and consider the two triangles thus formed.)

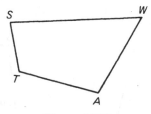

Figure 1.107

10. An **obtuse** angle is an angle which has more than 90°. Prove (in a sentence or two) that a triangle cannot have more than one obtuse angle. (You might try a proof by contradiction—suppose a triangle *could* have more than one.)

11. Can a triangle have *no* obtuse angles?

12. In the figure line M is perpendicular to line L,

$$m(\angle x) = m(\angle y) \qquad \text{and} \quad L \parallel K.$$

Prove that $m(\angle a) = m(\angle b)$.

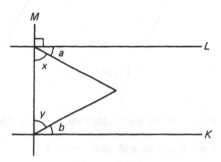

Figure 1.108

13. In the figure, $\overline{CL}$ is parallel to $\overline{AB}$. Prove that

$$m(\angle s) = m(\angle p) + m(\angle q).$$

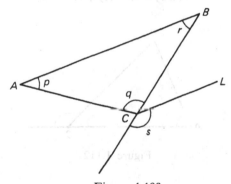

Figure 1.109

14. Let PMQ be a right triangle with right angle at M. Let the line through M perpendicular to $\overline{PQ}$ intersect $\overline{PQ}$ at a point H. Prove that the three angles of $\triangle PHM$ have the same measure as the three angles of $\triangle MHQ$.

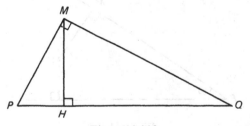

Figure 1.110

15. In the figure below, $TALK$ is a parallelogram. Prove that $m(\angle K) = m(\angle A)$ and that $m(\angle T) = m(\angle L)$.

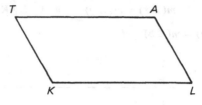

Figure 1.111

[*Hint*: Extend the sides of the parallelogram, and use Theorem 1-4.]

16. When you have done this, you will have proved:

Theorem 1-7. *The opposite angles of a parallelogram have the same measure.*

17. Let ABC be an arbitrary triangle, with $\overline{AC}$ extended. Prove that

$$m(\angle 1) = m(\angle A) + m(\angle B).$$

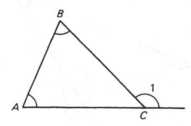

Figure 1.112

18. In the figure below, $\overline{QM}$ and $\overline{QN}$ are perpendicular to the rays defining angle A. Prove that angle A and angle C are supplementary.

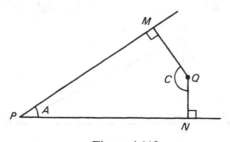

Figure 1.113

19. In the figure below, $\overline{QN}$ is perpendicular to $\overline{PS}$ and $\overline{QM}$ is perpendicular to $\overline{PT}$. Prove that angles $\angle P$ and $\angle Q$ have the same measure.

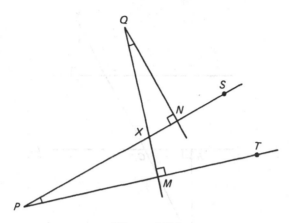

Figure 1.114

20. Figure *HELP* is a parallelogram. Prove that $\angle H$ is supplementary to $\angle P$. (In other words, that $m(\angle H) + m(\angle P) = 180°$.)

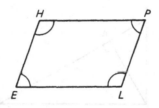

Figure 1.115

21. Given two lines L_1 and L_2 cut by a third line K. If $m(\angle A) = m(\angle B)$, prove that L_1 and L_2 are parallel. [*Hint*: Use Theorem 1-6.]

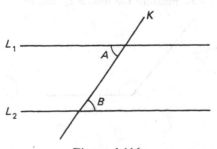

Figure 1.116

22. In the figure, $L_1 \parallel L_2$; $\overline{XY} \perp L_2$ and $\overline{AB} \perp L_1$. Prove

$$m(\angle A) = m(\angle X).$$

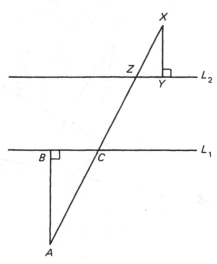

Figure 1.117

23. In right triangle ABC below, $\overline{CN}$ is drawn perpendicular to the hypotenuse. Prove that $m(\angle NCB) = m(\angle A)$.

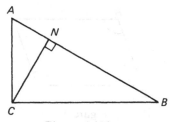

Figure 1.118

ADDITIONAL EXERCISES FOR CHAPTER 1

1. Using your protractors, measure the following angles:

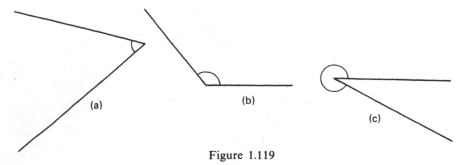

Figure 1.119

2. (Use ruler and compass.) Draw a triangle. Carefully bisect each angle of the triangle. What do you observe?

3. Circle those combinations of lengths which will form a triangle:
   (a) 2 cm, 2 cm, 2 cm   (b) 7 cm, 12 cm, 13 cm
   (c) 14 mm, 15 mm, 30 mm   (d) 2 km, 5 km, 3 km

4. Let P, Q, and M be points in the plane.
   (a) If $\angle QPM = 0°$, what can you conclude about P, Q, and M?
   (b) If $d(M, Q) = d(P, Q) + d(M, P)$ what can you conclude about P, Q, and M?

5. Explain briefly (using words and pictures) what's wrong with the following statement: "Points X and Y are in the disk of radius r around P if

$$d(X, Y) \leq 2r."$$

6. Without measuring, find x in each of the following:
   (In Figure (c), the horizontal lines are assumed parallel.)

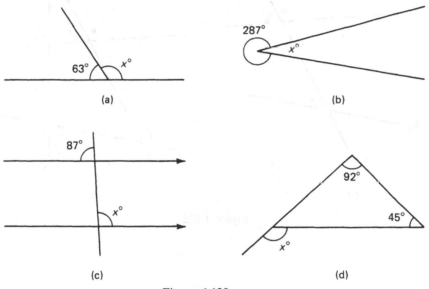

(a)

(b)

(c)

(d)

Figure 1.120

7. In the figure, L_1 is parallel to L_2. Prove that $\angle A$ is supplementary to $\angle B$.

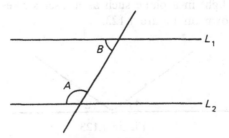

Figure 1.121

8. Prove that a right triangle cannot have an obtuse angle.

9. In the figure, L_1 and L_2 are any two lines. Line K intersects L_1 at P and L_2 at R. What's wrong with the following "proof"?

"From Q draw a line perpendicular to L_1 and L_2, meeting the lines at M and N. In right triangle QMP, we know by Theorem 1-3 that $m(\angle 1) + m(\angle Q) = 90°$. Similarly, in right triangle QNR, we know that $m(\angle 2) + m(\angle Q) = 90°$. Thus we obtain that

$$m(\angle 1) = m(\angle 2)."$$

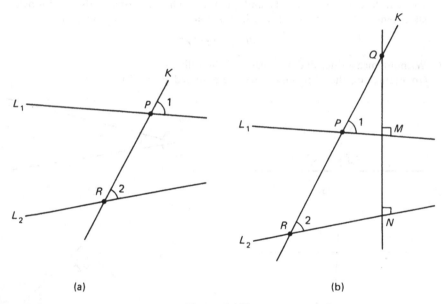

(a)   (b)

Figure 1.122

1, §7. AN APPLICATION: ANGLES OF REFLECTION

Suppose a ray of light in a plane such as a laser strikes a flat mirror and bounces off as shown on Figure 1.123.

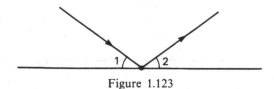

Figure 1.123

Then angle $\angle 1$ in the figure is called the **angle of incidence**, and angle $\angle 2$ is called the **angle of reflection**. All measurements which have been made confirm the law of physics that $m(\angle 1) = m(\angle 2)$, in other words:

The angle of incidence is equal to the angle of reflection.

Instead of representing a beam of light, the figure could also represent the path of a billiard ball which is shot straight (without spin) against the side of a billiard table. The same law applies.

Using results from the last section, prove the following beautiful property as an exercise.

Suppose a beam of light in the plane bounces off one side of a right angle, hits the other side, and again bounces off. Then the original path line of the beam is parallel to the path line of the second bounce. In other words, on Figure 1.124, the two segments $\overline{PQ}$ and $\overline{MN}$ are parallel.

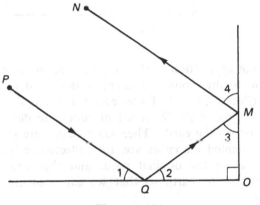

Figure 1.124

In the figure, $m(\angle O) = 90°$, so $\angle O$ is a right angle. Furthermore since the angle of incidence equals the angle of reflection, we have

$$m(\angle 1) = m(\angle 2) \qquad \text{and} \qquad m(\angle 3) = m(\angle 4).$$

From this, proving that $\overline{PQ}$ is parallel to $\overline{MN}$ is no harder than the exercises of §6. [*Hint for the proof*: Draw the figure with the intersection of the lines L_{PQ} and L_{MO}, and use Theorem 1-6.]

The same principle applies in a 3-dimensional situation. Suppose a beam of light is directed toward one side of a rectangular box, bounces off, hits another side, and again bounces off. Then the path line after the second bounce is parallel to the original path line, as illustrated on Figure 1.125. In other words,

$\overline{PQ}$ *is parallel to* $\overline{MN}$.

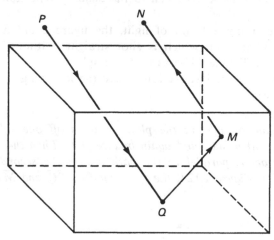

Figure 1.125

One interesting application of this principle occurs in measurements of laser beams sent to the moon and back, as described in the paper "The Lunar Laser Reflector", by J.E. Faller and E.J. Wampler, *Scientific American*, March 1970, pages 38-49. A lot of cubes are placed on the moon with openings toward the earth. Then laser beams are shot toward them. No matter how slanted the cubes are, the reflected path after the second bounce is parallel to the original path, and therefore the laser beam points back toward the earth so that we can measure the laser beam when it returns.

CHAPTER 2

Coordinates

2, §1. COORDINATE SYSTEMS

In more elementary courses you have already studied the number line, and so you already know that once a unit length is selected, we can represent points on a line by numbers. In Figure 2.1 we have drawn such a line and a few points.

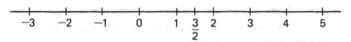

Figure 2.1

We shall now extend this procedure to the plane, and we shall represent points in the plane by pairs of numbers.

In Figure 2.2, we visualize a horizontal line and a vertical line intersecting at a point O, called the **origin**.

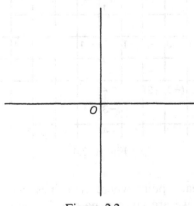

Figure 2.2

These lines will be called **coordinate axes**, or simply **axes**.

We select a unit length and cut the horizontal line into segments of lengths 1, 2, 3,... to the left and to the right. We do the same to the vertical line, but up and down, as indicated on Figure 2.3.

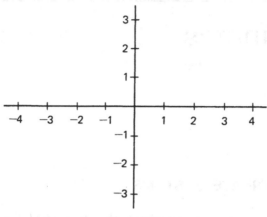

Figure 2.3

On the vertical line, we visualize the points going below O as corresponding to the negative integers, just as we visualize points on the left of the horizontal line as corresponding to negative integers. We follow the same idea as that used in grading a thermometer, where the numbers below zero are regarded as negative.

We can now cut the plane into squares whose sides have length 1.

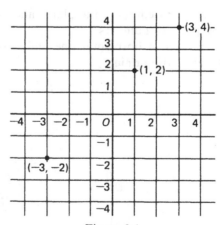

Figure 2.4

We can describe each point where two lines intersect by a pair of integers, like (1, 2). We go 1 unit to the right of the origin and up 2 units vertically to get the point

(1, 2) which has been indicated in Figure 2.4. We have also indicated the point (3, 4). The diagram is just like a map.

Furthermore, we could also use negative numbers. For instance, to describe the point $(-3, -2)$, we go 3 units to the left of the origin and 2 units vertically downward.

There is actually no reason why we should limit ourselves to points which are described by integers. For instance, we can also describe the point $(\frac{1}{2}, -1)$ and the point $(-\sqrt{2}, 3)$ as on Figure 2.5.

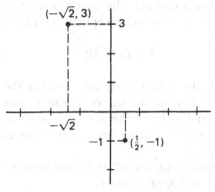

Figure 2.5

In general, if we take any point P in the plane and draw the perpendicular lines to the horizontal axis and to the vertical axis, we obtain two numbers x, y as on Figure 2.6.

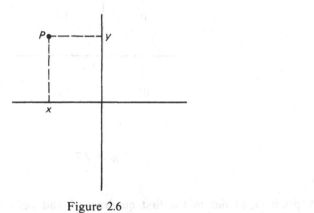

Figure 2.6

We define the numbers x, y to be the **coordinates** of the point P, and we write

$$P = (x, y).$$

Conversely, every pair of numbers (x, y) determines a point of the plane. If x is positive, then this point lies to the right of the vertical axis. If x is negative, then this point lies to the left of the vertical axis. If y is positive, then this point lies above the horizontal axis. If y is negative, then this point lies below the horizontal axis.

The coordinates of the **origin** are

$$O = (0, 0).$$

We usually call the horizontal axis the x-axis, and the vertical axis the y-axis (exceptions will always be noted explicitly). Thus if

$$P = (5, -10),$$

then we say that 5 is the x-coordinate and -10 is the y-coordinate. Of course, if we don't want to fix the use of x and y, then we say that 5 is the first coordinate, and -10 is the second coordinate. What matters here is the ordering of the coordinates, so that we can distinguish between a first and a second.

We can, and sometimes do, use other letters besides x and y for coordinates, for instance t and s, or u and v.

Our two axes separate the plane into four **quadrants**, which are numbered as indicated in Figure 2.7.

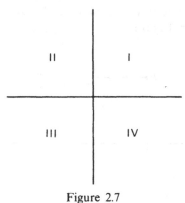

Figure 2.7

A point (x, y) lies in the **first quadrant** if and only if both x and y are > 0. A point (x, y) lies in the **fourth quadrant** if and only if $x > 0$, but $y < 0$.

Finally, we note that we placed our coordinates horizontally and vertically for convenience. We could also place the coordinates in a slanted way, as shown in Figure 2.8.

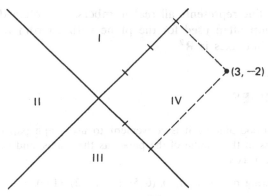

Figure 2.8

In Figure 2.3, we have indicated the quadrants corresponding to this coordinate system, and we have indicated the point $(3, -2)$ having coordinates 3 and -2 with respect to this coordinate system. Of course, when we change the coordinate system, we also change the coordinates of a point.

Remark. Throughout this book, when we select a coordinate system, the positive direction of the second axis will always be determined by rotating counterclockwise the positive direction of the first axis through a right angle.

We observe that the selection of a coordinate system amounts to the same procedure that is used in constructing a map. For instance, on the following (slightly distorted) map, the coordinates of Los Angeles are $(-6, -2)$, those of Chicago are $(3, 2)$, and those of New York are $(7.2, 3)$. (View the distortion in the same spirit as you view modern art.)

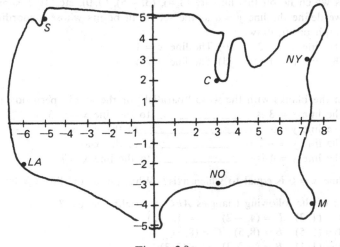

Figure 2.9

Since a number line represents all real numbers, it is often denoted by **R** or **R**^{1}. A notation often used for the plane with coordinates determined by two number line axes is **R**^{2}.

2, §1. EXERCISES

For exercises with coordinates, it is convenient to use graph paper. Choose one of the vertical lines in the middle of the paper as the y-axis, and one of the horizontal lines as the x-axis.

1. Plot the following points: $(-1, 1)$, $(0, 5)$, $(-5, -2)$, $(1, 0)$.

2. Plot the following points: $(\frac{1}{2}, 3)$, $(-\frac{1}{3}, -\frac{1}{2})$, $(\frac{4}{3}, -2)$, $(-\frac{1}{4}, -\frac{1}{2})$.

3. Let (x, y) be the coordinates of a point in the second quadrant. Is x positive or negative? Is y positive or negative?

4. Let (x, y) be the coordinates of a point in the third quadrant. Is x positive or negative? Is y positive or negative?

5. Plot the following points: $(1.2, -2.3)$, $(1.7, 3)$.

6. Plot the following points: $(-2.5, \frac{1}{3})$, $(-3.5, \frac{5}{4})$.

7. Plot the following points: $(1.5, -1)$, $(-1.5, -1)$.

8. What are the coordinates of Seattle (S), Miami (M) and New Orleans (NO) on our map? (See Figure 2.9.)

9. Tell whether the following points lie on the x-axis or the y-axis:
   (a) $(0, 4)$ (b) $(-4, 0)$ (c) $(-1, 0)$ (d) $(0, 0)$

10. Let a be a number. The line $x = a$ is defined as the set of all points whose x-coordinate is a. For example, if $a = 3$, we have the line $x = 3$. Some of the points which lie on this line are $(3, 4)$, $(3, -5)$, $(3, 0)$, etc. In a similar manner, we define the line $y = a$ as the set of all points whose y-coordinate is a. On graph paper, draw:
    (a) The line $x = -3$ (b) The line $x = 1$
    (c) The line $y = 0$ (d) The line $y = 4$

11. Fill in the blanks with the word "parallel" or the word "perpendicular".
    (a) The line $x = 3$ is _____ to the line $x = -3$.
    (b) The line $y = -7$ is _____ to the y-axis.
    (c) The line $y = -7$ is _____ to the x-axis.
    (d) The line $y = 4$ is _____ to the line $x = 7$.

12. The line $x = 0$ is equal to which axis? What about the line $y = 0$?

13. Which of the following triangles ABC are right triangles?
    (a) $A = (1, 3)$ $B = (4, -2)$ $C = (1, -2)$
    (b) $A = (5, 5)$ $B = (8, 5)$ $C = (8, 3)$
    (c) $A = (1, 1)$ $B = (-2, 3)$ $C = (3, 3)$

14. Which of the following triangles XYZ are isosceles triangles?
    (a) $X = (-4, -3)$ $Y = (-2, 2)$ $Z = (0, -3)$
    (b) $X = (1, -5)$ $Y = (2, -1)$ $Z = (9, -6)$
    (c) $X = (4, 0)$ $Y = (8, 2)$ $Z = (8, -2)$

15. Three vertices of a parallelogram are at $(0, 0)$, $(3, 5)$, and $(8, 0)$. What are the coordinates of the fourth vertex (there's more than one possibility)?

16. Plot some points whose x-coordinate and y-coordinate are equal. The set of all such points form what figure?

17. Plot some points whose y-coordinate is twice the x-coordinate. Describe the figure formed by all such points.

EXPERIMENT 2-1

Draw a number line on your paper such as the one below:

1. What is the distance between the points 3 and 5 on the line? Remember that distance is always a positive quantity.

2. What is the distance between the points -2 and 4 on the number line?

3. What is the distance between the points -3 and -1 on the number line?

2, §2. DISTANCE BETWEEN POINTS ON A LINE

First let us consider the distance between two points on a line. For instance, the distance between the points 1 and 4 on the line is $4 - 1 = 3$.

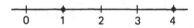

Observe that if we take the difference between 1 and 4 in the other direction, namely $1 - 4$, then we find -3, which is negative. However, if we then take the square, we find the same number, namely

$$(-3)^2 = 3^2 = 9.$$

Thus when we take the square, it does not matter in which order we took the difference.

Example. Find the square of the distance between the points -2 and 3 on the line.

The square of this distance is

$$(3 - (-2))^2 = (3 + 2)^2 = 25;$$

or computing the other way,

$$(-2 - 3)^2 = (-5)^2 = 25.$$

Note that again we take the difference between the coordinates of the points, and that we can deal with points having negative coordinates. If we want the distance rather than its square, then we take the square root, and we find

$$\sqrt{(-5)^2} = \sqrt{5^2} = \sqrt{25} = 5.$$

Because of our universal convention that the square root of a positive number is taken to be positive, we see that we can express the general formula for the distance between points on a line as follows.

Theorem 2-1. *Let* x_1, x_2 *be points on a number line. Then the distance between* x_1 *and* x_2 *is equal to*

$$\sqrt{(x_1 - x_2)^2}.$$

2, §2. EXERCISES

Find the distance between the following points on the number line.

1. 3 and 7

2. −3 and 7

3. 3 and −7

4. −3 and −7

5. 7 and 9

6. 7 and 10

7. −7 and −10

8. Let x be a number such that the distance from x to 5 is
   (a) 1   (b) 5   (c) 6   (d) 7
   What are the possible values of x?

9. Let x be a number such that the distance from x to −2 is
   (a) 2   (b) 3   (c) 4   (d) 5
   What are the possible values of x?

10. Let x be a number such that the distance from x to 8 is h. What are the possible values of x?

2, §3. EQUATION OF A LINE

In this section we shall describe lines in terms of coordinates. The section could be omitted without impairing the logical understanding of the book, but it would fit well with Chapter 12.

Let $F(x, y)$ be an expression involving a pair of numbers (x, y). Let c be a number. We consider the equation

$$F(x, y) = c.$$

We define the **graph** of the equation to be the set of points (x, y) such that $F(x, y) = c$. In this section, we study the simple case of equations like

$$3x - 2y = 5.$$

Similarly, if $y = F(x)$ is given as an expression in x, we define the **graph of the curve whose equation is $y = F(x)$** to be the set of points $(x, F(x))$.

Consider first a simple example, namely

$$y = 3x.$$

The set of points $(x, 3x)$ is the graph of this equation, or equivalently of the equation

$$y - 3x = 0.$$

We can give x an arbitrary value, and thus we see that the graph looks like Figure 2.10(a).

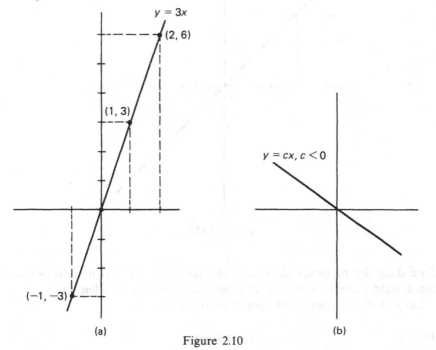

(a)　　　　　　　　　　　　　　　　　　　　(b)

Figure 2.10

If we consider the graph of

$$y = 4x,$$

we see that it is a line which slants more steeply. In general, the graph of the equation

$$y = ax,$$

where a is a number, represents a straight line. An arbitrary point on this line is of type

$$(x, ax).$$

If a is positive, then the line slants to the right. If a is negative, then the line slants to the left, as shown in Figure 2.10(b). For instance, the graph of

$$y = -x$$

consists of all points $(x, -x)$, and looks like Figure 2.11:

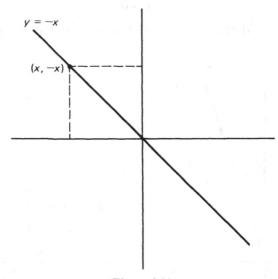

Figure 2.11

If we drop the perpendiculars from the point $(x, -x)$ to the axis, we obtain a right triangle, which in this case has legs of equal length.

Let a, b be numbers. The graph of the equation

(1) $$y = ax + b$$

is also a straight line, which is parallel to the graph of the equation

(2) $y = ax.$

To convince ourselves of this, we observe the following. Let

$$y' = y - b.$$

The equation

(3) $y' = ax$

is of the type just discussed. If we have a point (x, y') on the graph of (3), then we get a point $(x, y' + b)$ on the graph of (1), simply by adding b to the second coordinate. This means that the graph of the equation

$$y = ax + b$$

is the straight line parallel to the line determined by the equation

$$y = ax,$$

and passing through the point $(0, b)$.

Example. We want to draw the graph of the equation

$$y = 2x + 1.$$

When $x = 0$, then $y = 1$. When $y = 0$, then $x = -\frac{1}{2}$. Hence the graph looks like Figure 2.12(b).

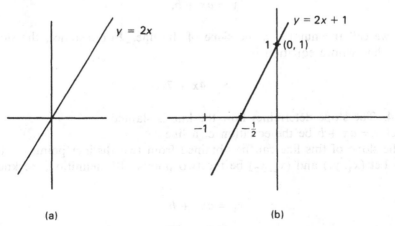

(a)   (b)

Figure 2.12

Example. We want to draw the graph of the equation

$$y = -2x - 5.$$

When $x = 0$, then $y = -5$. When $y = 0$, then $x = -\frac{5}{2}$. Hence the graph looks like Figure 2.13.

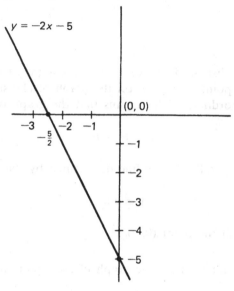

Figure 2.13

If a line L is the graph of the equation

$$y = ax + b,$$

then we call the number a the **slope** of the line. For instance, the slope of the line whose equation is

$$y = -4x + 7$$

is -4. The slope determines how the line is slanted.

Let $y = ax + b$ be the equation of a line.

The slope of this line can be obtained from two distinct points on the line. Let (x_1, y_1) and (x_2, y_2) be the two points. By definition, we know that

$$y_1 = ax_1 + b,$$

$$y_2 = ax_2 + b.$$

Subtracting, we find that

$$y_2 - y_1 = a(x_2 - x_1).$$

Consequently, if $x_2 \neq x_1$, we can divide by $(x_2 - x_1)$ to find

$$\text{slope} = a = \frac{y_2 - y_1}{x_2 - x_1}.$$

This formula gives us the slope in terms of the coordinates of two distinct points.

Example. Consider the line defined by the equation

$$y = 2x + 5.$$

The two points $(1, 7)$ and $(-1, 3)$ lie on the line. The slope is equal to 2, and, in fact,

$$2 = \frac{7 - 3}{1 - (-1)},$$

as it should be.

Geometrically, our quotient

$$\frac{y_2 - y_1}{x_2 - x_1}$$

is the ratio of the vertical side and horizontal side of the triangle in Figure 2.14.

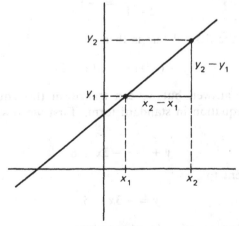

Figure 2.14

Observe that it does not matter which point we call (x_1, y_1) and which one we call (x_2, y_2). We would get the same value for the sope. This is because if we invert their order, then in the quotient expressing the slope both the numerator and denominator will change by a sign, so that their quotient does not change.

In general, the equation of the line passing through the point (x_1, y_1) and having slope a is

$$y - y_1 = a(x - x_1).$$

For points such that $x \neq x_1$, we can also write this in the form

$$\frac{y - y_1}{x - x_1} = a.$$

The equation of the line passing through two points (x_1, y_1) and (x_2, y_2) is

$$\frac{y - y_1}{x - x_1} = \frac{y_2 - y_1}{x_2 - x_1}$$

for all points such that $x \neq x_1$.

Example. Find the equation of the line passing through the points $(1, 2)$ and $(2, -1)$.

We first find the slope. It must be the quotient

$$\frac{y_2 - y_1}{x_2 - x_1},$$

which in this case is equal to

$$\frac{-1 - 2}{2 - 1} = -3.$$

The equation of the desired line is then

$$y - (-1) = -3(x - 2).$$

This is a correct answer, but we can transform this equation by simple algebra into an equation in standard form. First we rewrite the equation as

$$y + 1 = -3x + 6,$$

which is equivalent to

$$y = -3x + 5.$$

This is the simplest equation for the desired line.

Example. Find the equation of the line having slope -7 and passing through the point $(-1, 2)$.

The equation has the form

$$y - 2 = -7(x - (-1)).$$

By simple algebra, this is equivalent with

$$y - 2 = -7(x + 1) = -7x - 7.$$

Again, this is equivalent with

$$y = -7x - 5.$$

This is the equation in standard form.

We should also mention vertical lines. These cannot be represented by equations of type

$$y = ax + b.$$

Suppose that we have a vertical line intersecting the x-axis at the point $(2, 0)$. The y-coordinate of any point on the line can be an arbitrary number, while the x-coordinate is always 2. Hence the equation of this line is simply

$$x = 2.$$

Similarly, the equation of a vertical line intersecting the x-axis at the point $(c, 0)$ is

$$x = c.$$

2, §3. EXERCISES

Sketch the graphs of the following lines.

1. (a) $y = -2x + 5$ (b) $y = 5x - 3$

2. (a) $y = \dfrac{x}{2} + 7$ (b) $y = -\dfrac{x}{3} + 1$

3. (a) $y = 3x - 1$ (b) $y = -4x + 2$

What is the equation of the line passing through the fol'owing points?

4. (a) $(-1, 1)$ and $(2, -7)$ (b) $(3, \frac{1}{2})$ and $(4, -1)$

5. (a) $(\sqrt{2}, -1)$ and $(\sqrt{2}, 1)$ (b) $(-3, -5)$ and $(\sqrt{3}, 4)$

What is the equation of the line having the given slope and passing through the given point?

6. Slope 4 and point $(1, 1)$

7. Slope -2 and point $(\frac{1}{2}, 1)$

8. Slope $-\frac{1}{2}$ and point $(\sqrt{2}, 3)$

9. Slope 3 and point $(-1, 5)$

Sketch the graphs of the following lines:

10. $x = 5$

11. $x = -1$

12. $x = -3$

13. $y = -4$

14. $y = 2$

15. $y = 0$

What is the slope of the line passing through the following points?

16. $(1, \frac{1}{2})$ and $(-1, 1)$

17. $(\frac{1}{4}, 1)$ and $(\frac{1}{2}, -1)$

18. $(2, 1)$ and $(\sqrt{2}, 1)$

19. $(\sqrt{3}, 1)$ and $(3, 2)$

What is the equation of the line passing through the following points?

20. $(\pi, 1)$ and $(\sqrt{2}, 3)$

21. $(\sqrt{2}, 2)$ and $(1, \pi)$

22. $(-1, 2)$ and $(\sqrt{2}, -1)$

23. $(-1, \sqrt{2})$ and $(-2, -3)$

24. Sketch the graphs of the following lines.
    (a) $y = 2x$ (b) $y = 2x + 1$ (c) $y = 2x + 5$
    (d) $y = 2x - 1$ (e) $y = 2x - 5$

CHAPTER 3

Area and the Pythagoras Theorem

3, §1. THE AREA OF A TRIANGLE

We have already relied on your intuitive notion of area in Postulate **RT**, when we spoke of two right triangles "having the same area". In the first section, we show how to compute areas of various figures.

Whenever we choose a unit of distance to measure the length of a line segment (as we did in the first chapter), we are also selecting a unit of area, with which we can measure the amount of space enclosed by a region in the plane. For example, if we select the centimeter as the unit of distance, the corresponding unit of area is the square centimeter. One **square centimeter** is defined as the area of the region enclosed by a square with sides of length one centimeter:

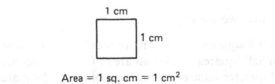

Area = 1 sq. cm = 1 cm²

Similarly, one square meter is the area of the region enclosed by a square with sides of length one meter. Other "unit areas" can be defined using other units of length.

EXPERIMENT 3-1 (Omit if familiar with basic area formulas.)

We can find the **area** of various regions in the plane by determining how many unit area squares fit into the region. For example, suppose ·we have a square with sides of length 4 cm; we see that its area is 16 square cm:

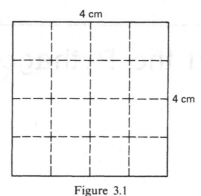

Figure 3.1

Suppose the square had sides of length $3\frac{1}{2}$ cm. Look at the picture, and count the number of unit squares, adding up the fractional parts of squares:

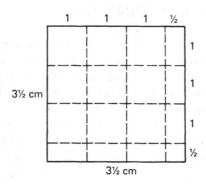

Figure 3.2

You will see that we have:

9 full squares	for an area of	9 square cm
6 half squares	for an area of	3 square cm
1 quarter square	for an area of	$\frac{1}{4}$ square cm

TOTAL AREA $12\frac{1}{4}$ square cm

1. Draw a picture illustrating a square whose sides have length $10\frac{1}{4}$ cm. Determine its area by counting unit squares and fractions of unit squares.

2. Draw a rectangle whose sides have lengths 3 cm and 5 cm. Determine its area by counting unit squares.

3. Draw a rectangle whose sides have lengths $9\frac{1}{4}$ cm and 5 cm. Count squares to determine its area.

4. In your picture for Problem 2, draw a **diagonal** of the rectangle (a line segment joining opposite corners), which creates two right triangles.
   (a) How do the areas of these two triangles compare? Why?
   (b) Compute the areas of these triangles without counting.

The general formulas giving the areas of a square and a rectangle are as follows.

Area of a square whose sides have length s is

$$s \cdot s = s^2 \quad square \ units.$$

Area of a rectangle whose sides have lengths a, b units is

$$ab \quad square \ units.$$

These formulas will be taken for granted, but the experiment should convince you of their truth.

Example. Look back at Part 3. To compute the area, we write

$$\text{area} = 9\frac{1}{4} \cdot 5 = 46\frac{1}{4} \text{ square cm.}$$

Now rewrite the multiplication as

$$(9 + \tfrac{1}{4}) \cdot 5 = 9 \cdot 5 + \tfrac{1}{4} \cdot 5.$$

Looking at the right-hand side of this equation, see if you can discern the relationship between it and the actual number of squares in the rectangle.

Use a similar procedure to show that the number of unit squares (adding up fractional parts when necessary) in a rectangle with sides of length $2\frac{1}{2}$ inches and $3\frac{1}{2}$ inches can be found by multiplying the number $2\frac{1}{2}$ by the number $3\frac{1}{2}$.

5. Using whatever method you wish, calculate the area of the figure given below:

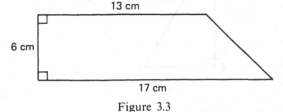

Figure 3.3

A general formula for the area of such a figure will be derived in Theorem 3-3, but think first before you look it up.

Hopefully you saw in the experiment that the familiar formulas for the areas of squares and rectangles merely calculate the number of *units* of area in each region, and therefore they determine what we intuitively call area, rather than appear simply as formulas to be memorized.

Incidentally, the lengths of sides of these figures need not only be whole numbers or fractions; they may be numbers such as $\sqrt{2}$ or π or any other positive irrational number.

Example. The area of the rectangle illustrated below is $\sqrt{6}$ sq. cm.

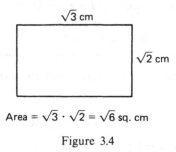

$\sqrt{3}$ cm

$\sqrt{2}$ cm

Area = $\sqrt{3} \cdot \sqrt{2} = \sqrt{6}$ sq. cm

Figure 3.4

From Part 4 of the experiment, you may have already discovered the following theorem:

Theorem 3-1. *The area of a right triangle is one-half the product of the lengths of the two legs.*

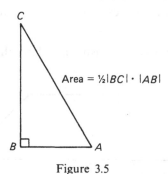

C

Area = $\frac{1}{2}|BC| \cdot |AB|$

B A

Figure 3.5

Proof. Given triangle *ABC*, we then construct a rectangle *PQRS* whose sides have the same lengths as $\overline{AB}$ and $\overline{BC}$, and we draw diagonal $\overline{QS}$:

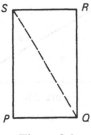

Figure 3.6

By our postulate **RT**, triangles $\triangle QPS$, $\triangle SRQ$, and $\triangle ABC$ have the same area. We know:

$$\text{area of rectangle } PQRS = |PQ||RS|$$
$$= |AB||BC|.$$

The area of one of the two (equal-area) triangles in the rectangle then must be equal to

$$\tfrac{1}{2}|AB||BC|,$$

which then also must be the area of $\triangle ABC$. This proves the theorem.

Example. The area of the right triangle in the figure is equal to

$$\text{area} = \tfrac{1}{2} \cdot 6 \cdot 5 = 15 \text{ sq. cm.}$$

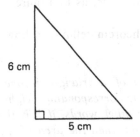

6 cm

5 cm

Figure 3.7

Example. The area of the right triangle in the figure is given by:

$$\text{area} = \tfrac{1}{2} \cdot 2a \cdot a = a^2 \text{ square units.}$$

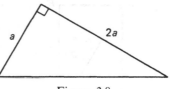

Figure 3.8

We also use Theorem 3-1 to prove a theorem which tells us the area of *any* triangle, not just right triangles. First, we define a couple of new terms to simplify our discussion. Any side of a triangle can be considered a **base** of the triangle. We choose such a base. The perpendicular segment from the opposite vertex to the line of the base is called the **height** or the **altitude** of the triangle relative to that base. Sometimes the **length of that segment** is also called the **height**. In Figure 3.9, we have illustrated two possibilities for the base and height of a triangle *ABC*.

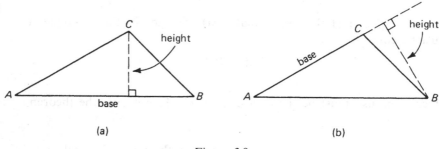

Figure 3.9

Notice that the segment which gives us the height relative to a given base may fall outside the triangle, as in Figure 3.9(b).

We can now prove a theorem telling us how to compute the area of *any* triangle.

Theorem 3-2. *The area of a triangle is one-half the product of the lengths of the base and its corresponding height, regardless of which side we choose as a base. In other words, if b is the length of a base, and h is the corresponding height, then the area of the triangle is*

$$\text{area} = \tfrac{1}{2}bh.$$

Proof. There are two cases to consider, depending on whether the height segment falls inside the triangle or outside the triangle.

Case I. The height falls inside the triangle and divides the base into two parts, with lengths x and y:

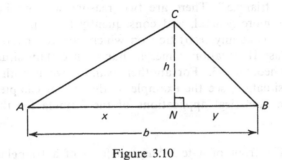

Figure 3.10

We see that the area of our triangle ABC is equal to:

$$\text{area of } \triangle ANC + \text{area of } \triangle BNC.$$

Theorem 3-1 tells us how to compute the areas of these two right triangles:

$$\text{area } \triangle ANC = \tfrac{1}{2}xh,$$

$$\text{area } \triangle BNC = \tfrac{1}{2}yh.$$

Therefore,

$$\text{area of } \triangle ABC = \tfrac{1}{2}xh + \tfrac{1}{2}yh$$

$$= \tfrac{1}{2}h(x + y)$$

$$= \tfrac{1}{2}hb,$$

since $x + y = b$.

The second case is illustrated below:

Case II.

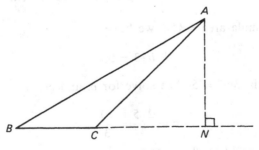

Figure 3.11

The proof is left as an exercise for you. Notice that the area of $\triangle ABC$ is equal to the area of right triangle ABN *minus* the area of right triangle ACN.

Students sometimes ask, "Why bother with Case II? Whenever you need to find the area of a triangle, just choose a base so that the height falls inside the triangle." There are two reasons why we bother. First, the theorem is more general, and consequently more pleasing, when we do not have to make any restriction on which side you choose as a base. Second, the Case II situation comes in handy in certain situations, as in the proof of Theorem 3-3. For another example when both cases occur in a physical situation, see the example at the end of Chapter 7, §1. We shall now give numerical applications of the formula for the area of a triangle.

Example. The front of a tent has the form of a triangle, and has a base of length 1.5 m and a height of 2 m. How much material is needed to make a cover for this front?

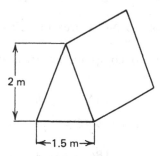

Figure 3.12

By our general formula, the area is equal to

$$\tfrac{1}{2} \cdot 1.5 \cdot 2 = 1.5 \text{ m}^2.$$

This is the answer.

Example. Find the base of a triangle if its area is 5 cm² and its height is 3 cm.

From the formula area $= \tfrac{1}{2}bh$, we have

$$\tfrac{1}{2} \cdot b \cdot 3 = 5,$$

or in other words, $3b/2 = 5$. We solve for b, and get

$$b = \frac{2 \cdot 5}{3} = \frac{10}{3}.$$

Hence the answer is $b = \tfrac{10}{3}$ cm.

Trapezoids

Another figure which comes up in practical as well as theoretical situations is the trapezoid. A **trapezoid** is a four-sided figure which has a pair of parallel sides, called the **bases**. Figure 3.13 illustrates some trapezoids:

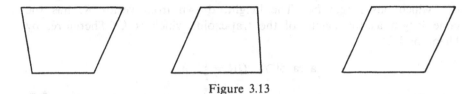

Figure 3.13

A property lot, or a water basin may have the shape of a trapezoid (see Exercises 17 and 18). It is important to have a formula for its area, which we shall now investigate.

The **height of the trapezoid** by definition is the perpendicular distance between the parallel lines going through the bases. We illustrate the height in the next figure.

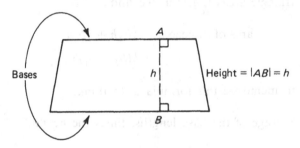

Figure 3.14

Theorem 3-3. *The area of a trapezoid with bases of length b_1 and b_2 and height h is:*

$$\tfrac{1}{2}(b_1 + b_2)h.$$

Proof. Given trapezoid $PQRS$, with lengths as indicated:

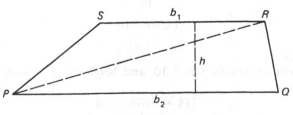

Figure 3.15

We draw a diagonal $\overline{PR}$, and notice that:

$$\text{area of trapezoid} = \text{area of } \triangle PQR + \text{area of } \triangle PSR.$$

So all we need to do is find the areas of the two triangles.

Looking at triangle PQR, the obvious choice for a base is segment $\overline{PQ}$, which has length b_2. The height, drawn from vertex R, has the same length as the height of the trapezoid, which is h. Therefore, by Theorem 3-2:

$$\text{area of } \triangle PQR = \tfrac{1}{2}b_2 h.$$

Looking at triangle PSR, the obvious choice for base is segment $\overline{SR}$, which has length b_1. The height drawn from vertex P to (an extension of) the base $\overline{SR}$ has the same length again as the height of the trapezoid, namely h. Therefore, by Theorem 3-2 (and here's where Case II comes in handy)

$$\text{area of } \triangle PSR = \tfrac{1}{2}b_1 h.$$

Adding these triangle areas together, we find

$$\text{area of trapezoid} = \tfrac{1}{2}b_2 h + \tfrac{1}{2}b_1 h$$
$$= \tfrac{1}{2}(b_2 + b_1)h.$$

An easy way to memorize this formula is to think:

"Average of the base lengths, times the height."

Example. Find the area of the trapezoid shown below:

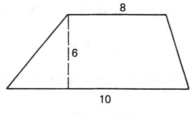

Figure 3.16

We have bases of lengths 8 and 10, and height 6. The area is therefore

$$\tfrac{1}{2}(8 + 10) \cdot 6 = 54.$$

3, §1. EXERCISES

1. In each of the following, the base b and height h of a triangle is given. Find the area:

   (a) $b = 12$ cm, $h = 5$ cm   (b) $b = 3$ m, $h = 7$ m   (c) $b = 2x$, $h = x$

2. Find a side of a square whose area is equal to the area of a rectangle with sides 10 and 40.

3. The height of a triangle is one half the base. Find its base if the area of the triangle is 36 sq. m.

4. Find the dimensions of a rectangle whose length is five times its width and whose area is 1440 sq. cm.

5. The piece of sheet metal illustrated below folds into a box with a square bottom and no top:

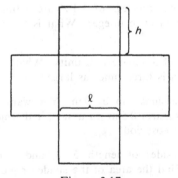

Figure 3.17

   (a) How much material is needed (in square cm) if $l = 5$ cm and $h = 4$ cm?

   (b) Write a formula for the area of the metal in terms of l and h.

6. A man wishes to build a path 1 m wide around a garden which is a rectangle 15 m by 20 m. What is the area of the path?

7. The sides of a rectangle are in the ratio 3:4 and its area is 300 sq. cm. Find its sides.

8. Find the base of a triangle if its area is 36 sq. cm and its height is 12 cm.

9. A right triangle has legs of equal length and area 40. How long are the legs?

10. In right triangles $\triangle ABC$ and $\triangle XYZ$ below, each leg in $\triangle XYZ$ has twice the length of the corresponding leg of $\triangle ABC$. What is the ratio of the area of $\triangle ABC$ to the area of $\triangle XYZ$?

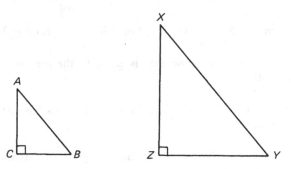

Figure 3.18

[*Hint*: Let a, b be the lengths of $\overline{AC}$ and $\overline{BC}$. Then $\overline{XZ}$ and $\overline{ZY}$ have lengths $2a$ and $2b$. Compute the areas and compare.]

11. In Problem 10, suppose the legs of XYZ are n times as big as the legs in $\triangle ABC$, where n is a positive integer. What is the ratio of the areas of the triangles?

12. The length of a side of a square is 8 units. What is the length of a side of the square whose area is three times as large?

13. A swimming pool measures 6 m by 9 m. You want to cover the floor with tiles which come in squares 0.5 m on a side, and which cost $35 each. How much will this project cost you?

14. Three squares, with sides of length 5, 4, and 3 respectively, are placed together as shown. Find the area of the shaded region.

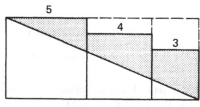

Figure 3.19

[*Hint*: Find this area by subtraction.]

15. Prove Case II of Theorem 3-2. In this case, the area of the given triangle is a *difference* of areas of two right triangles.

16. If a trapezoid has a base 8 and height 7, what is the length of the other base if the area of the trapezoid is 77.

17. A swimming pool is 30 meters long and 1 meter deep at the shallow end. The deep end extends for 4 meters at a depth of 5 meters. What is the area of the illustrated side wall?

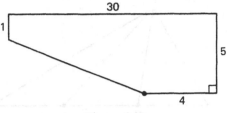

Figure 3.20

[*Hint*: Divide the region into a rectangle and a trapezoid.]

18. Three city lots, each with 25 m frontage along the main drag, make up a single tract, as illustrated below:

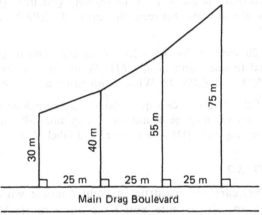

Figure 3.21

What is the total area of the tract?

19. Let $\triangle PQM$ be a triangle, as shown below. Let N be the midpoint of the segment $\overline{QM}$. Prove that the triangles $\triangle PQN$ and $\triangle PNM$ have the same area.

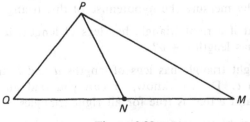

Figure 3.22

[*Hint*: What is the relation between the heights of these two triangles? Their bases?]

20. Let $\triangle PQM$ be a triangle, as shown below. Let N be the point on the segment $\overline{QM}$ such that the distance from N to M is twice the distance from Q to N, as shown.

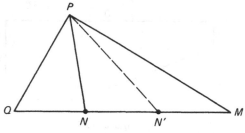

Figure 3.23

(a) Prove that the triangle $\triangle PNM$ has twice the area as $\triangle PQN$.

(b) How does the area of $\triangle PN'M$ relate to the area of $\triangle PQN$ if N' is the point two thirds of the way on the segment $\overline{QM}$ from Q to M?

(c) What is the relation between the area of $\triangle PNN'$ and the area of $\triangle PQN$?

21. In Exercise 20 suppose that we select N on the segment $\overline{QM}$ at a distance from Q equal to one-fourth of $d(Q, M)$. What can you then say about the areas of $\triangle PQN$ and $\triangle PNM$? What if this distance was one-fifth?

22. Prove that if the diagonals of a quadrilateral are perpendicular, then the area of the quadrilateral may be found by taking one-half the product of the lengths of the diagonals. (Draw a picture, and label all lengths clearly.)

EXPERIMENT 3-2

1. Using a ruler, carefully construct a right triangle with legs of length 8 cm and 13 cm. To be sure you have an accurate 90° angle, use a protractor, the corner of a piece of cardboard, or the procedure given in Construction 4 of Chapter 1.

2. Carefully measure the length of the hypotenuse of the triangle you constructed.

3. Construct another right triangle with legs of length 10 cm and 10 cm. Again carefully measure the hypotenuse of this triangle.

4. Is it true that if a right triangle has legs of lengths a and b, then the hypotenuse has length $a + b$?

5. Suppose a right triangle has legs of lengths a and b and the hypotenuse is length c. Do you know, or can you find an equation which relates a, b, and c that is true for all right triangles?

6. Construct a right triangle with legs of lengths 3 cm and 4 cm. Measure the hypotenuse.

7. Construct a right triangle with legs of lengths 8 and 15. Measure the hypotenuse.

3, §2. THE PYTHAGORAS THEOREM

In the previous Experiment, you investigated the relationship among the lengths of the sides of a right triangle. Suppose a right triangle has legs of lengths x and y, and hypotenuse of length z. The relation $x + y = z$ is false. The correct relation will be stated and proved below, and is generally attributed to Pythagoras. He was the leader of a group of philosophers which flourished for over 100 years around 530 B.C. The group was interested in music, morals, and religion as well as mathematics. Their contributions to geometry include proofs that the angles of a triangle contain 180°, work on the theory of parallels, and on proportions (similarities, as we would say today).

Theorem 3-4 (Pythagoras). *Let $\triangle XYZ$ be a right triangle with legs of lengths x and y, and hypotenuse of length z. Then*

$$x^2 + y^2 = z^2.$$

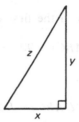

Figure 3.24

Proof. Let $PQRS$ be a square with sides of length $x + y$, and locate points A, B, C, D on the sides, such that each side is divided into two segments, one of length x and one of length y, as shown.

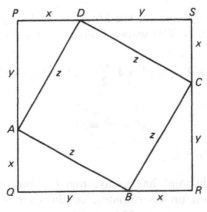

Figure 3.25

Connecting points A, B, C, D with four segments $\overline{AB}$, $\overline{BC}$, $\overline{CD}$, $\overline{DA}$ creates four right triangles to which we can apply postulate **RT**. Consequently, the lengths of their hypotenuses are all the same, namely z, and their areas are equal.

We now want to show that $ABCD$ is a square. First, it clearly does have four sides of equal length z. As for the angles, we start by looking at $\angle ABC$, and we see that:

$$m(\angle ABQ) + m(\angle ABC) + m(\angle CBR) = 180°.$$

By **RT**,
$$m(\angle CBR) = m(\angle QAB)$$

and by Theorem 1-3, we have

$$m(\angle QAB) + m(\angle ABQ) = 90°.$$

Therefore

$$m(\angle CBR) + m(\angle ABQ) = 90°.$$

Substituting these quantities in the first equation, we have

$$m(\angle ABC) + 90° = 180°$$

and therefore we conclude that:

$$m(\angle ABC) = 90°.$$

A similar argument holds for the other three angles in $ABCD$, and so we conclude that it is a square.

We now compute areas. Since the length of one side of square $PQRS$ is $x + y$, the total area enclosed by $PQRS$ is:

$$(x + y)^2 = x^2 + 2xy + y^2.$$

This area is equal to the sum of the areas of the four right triangles, plus the area of the square $ABCD$ whose sides have length z. Therefore

$$x^2 + 2xy + y^2 = 4\,\frac{xy}{2} + z^2 = 2xy + z^2,$$

which simplifies to

$$x^2 + y^2 = z^2,$$

and the theorem is proved.

The Pythagoras theorem has many, many applications (see the next chapter, or the section on coordinates, or the exercises). Consequently, you should learn it well, and be able to use it easily. Here are some typical calculations.

Example. Find the length of the diagonal of a square whose sides have length 1.

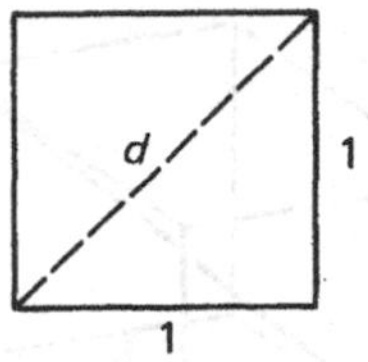

Figure 3.26

Label this length d. We know that $1^2 + 1^2 = d^2$ and so

$$d = \sqrt{1^2 + 1^2} = \sqrt{2}.$$

Example. What is the length of the diagonal of a rectangle whose sides have lengths 3 and 4? Here we have $d^2 = 3^2 + 4^2$, so that

$$d = \sqrt{9 + 16} = 5$$

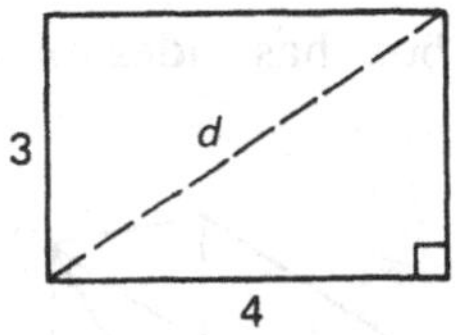

Figure 3.27

Example. One leg of a right triangle has length 10 cm, and the hypotenuse has length 15 cm. What is the length of the other side? Let b be this length. By Pythagoras we have

$$10^2 + b^2 = 15^2$$

Therefore

$$b^2 = 15^2 - 10^2,$$

$$b = \sqrt{225 - 100},$$

$$b = \sqrt{125} \text{ cm.}$$

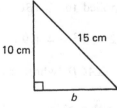

Figure 3.28

The Pythagoras theorem is also useful in 3-dimensional situations. Consider an ordinary open box. Suppose we wish to calculate the length of the diagonal from one corner of the box to the completely opposite corner as illustrated:

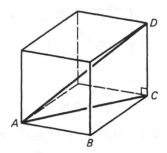

Figure 3.29

Observe that $\triangle ABC$ and $\triangle ACD$ are right triangles, with right angles at B and C respectively. If we are given the three sides of the box, we can use the Pythagoras theorem twice to compute the diagonal.

Example. A rectangular box has sides of lengths 2, 3, 7. Find the length of the diagonal.

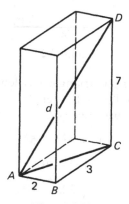

Figure 3.30

By the Pythagoras theorem, applied to $\triangle ABC$, we have

$$|AC|^2 = 2^2 + 3^2 = 4 + 9 = 13.$$

Next we apply Pythagoras to $\triangle ACD$, whose right angle is at C. Then we have the relation

$$|AC|^2 + 7^2 = d^2,$$

where d is the length of the diagonal of the box. Therefore

$$d^2 = 13 + 7^2 = 13 + 49 = 62.$$

Finally, we find $d = \sqrt{62}$.

Example. A right triangle has legs of lengths 6 and 7 units.

(a) Compute the length of the hypotenuse.
(b) If the legs are multiplied by a factor of 2, how does the hypotenuse change?
(c) If the legs are multiplied by a factor of r for some positive number r, how does the length of the hypotenuse change?

First, for (a), the hypotenuse has length

$$\sqrt{6^2 + 7^2} = \sqrt{36 + 49} = \sqrt{85}.$$

If the legs are multiplied by a factor of 2, so the new legs have lengths 12 and 14 respectively, the hypotenuse of the new triangle has length

$$\sqrt{12^2 + 14^2}.$$

The numbers are getting large, and so it is better to write this expression without performing the squaring operations yet, in the form

$$\sqrt{(2 \cdot 6)^2 + (2 \cdot 7)^2} = \sqrt{2^2 6^2 + 2^2 7^2}$$
$$= \sqrt{2^2(6^2 + 7^2)}.$$

We know a rule of algebra which says that

$$\sqrt{ab} = \sqrt{a}\sqrt{b}.$$

Hence we conclude that

$$\sqrt{(2 \cdot 6)^2 + (2 \cdot 7)^2} = \sqrt{2^2 \cdot 85} = \sqrt{2^2}\sqrt{85} = 2\sqrt{85}.$$

Thus we see that the hypotenuse of the new triangle differs from $\sqrt{85}$ by a factor of 2.

In general, if the legs are multiplied by a factor of r, then the hypotenuse changes to

$$\sqrt{(6r)^2 + (7r)^2} = \sqrt{r^2 6^2 + r^2 7^2} = \sqrt{r^2 85} = r\sqrt{85}.$$

Hence the hypotenuse changes by a factor of r.

Example. The lengths of the legs of a right triangle are in the ratio $3:5$. If the area of the triangle is 16, how long is the hypotenuse?

We can write the lengths of the legs in the form $3a$ and $5a$ for some number a. Then the area formula gives

$$16 = \frac{1}{2}\, 3a5a = \frac{15}{2}\, a^2.$$

Thus by algebra,

$$a^2 = \frac{32}{15}.$$

By Pythagoras, the hypotenuse squared is given by the formula

$$z^2 = (3a)^2 + (5a)^2 = 9a^2 + 25a^2 = 34a^2.$$

Substituting the value for a^2 which we found above, we obtain

$$z^2 = 34 \cdot \frac{32}{15}.$$

Hence

$$z = \sqrt{\frac{34 \cdot 32}{15}}.$$

This is a correct answer, and there is no particular need to simplify it.

Theorem 3-5. *Let P be a point and L a line. Suppose P is not on the line. The shortest distance between P and points on the line is the distance $d(P, M)$ where $\overline{PM}$ is the perpendicular segment from P to the line.*

Proof. Let Q be any point on the line unequal to M. Then P, M, Q are the three vertices of a right triangle, with right angle at M, as shown on the figure.

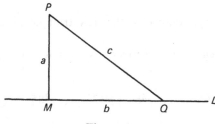

Figure 3.31

By Pythagoras, we have

$$a^2 + b^2 = c^2,$$

and since $b > 0$ it follows that $b^2 > 0$ and hence $c^2 > a^2$. Hence $c > a$. This means that

$$d(P, Q) = c > a = d(P, M),$$

thus proving the theorem.

3, §2. EXERCISES

1. Find the length x in each of the following diagrams:

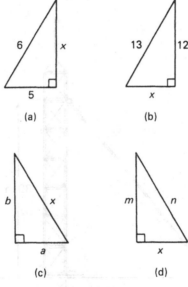

Figure 3.32

2. What is the length of the diagonal of a square whose sides have length:
   (a) 2   (b) 3   (c) 4   (d) 5   (e) r?

3. What is the length of the diagonal of a rectangle whose sides have lengths:
   (a) 1 and 2   (b) 3 and 5   (c) 4 and 7   (d) r and $2r$
   (e) $3r$ and $5r$ (f) $4r$ and $7r$?

4. What is the length of the diagonal of a cube whose sides have length:
   (a) 1   (b) 2   (c) 3   (d) 4   (e) r?

5. Find the length of the diagonal of a rectangular solid whose sides *a*, *b*, and *c* have lengths:

   (a) 3, 4, 5   (b) 1, 2, 4   (c) 1, 3, 4   (d) *a, b, c*   (e) *ra, rb, rc*

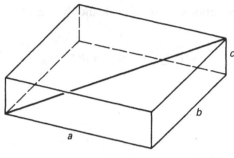

Figure 3.33

6. A TV tower is to be anchored by wires from a point 30 m above the ground to stakes set 25 m from the base of the tower. How long does each wire have to be?

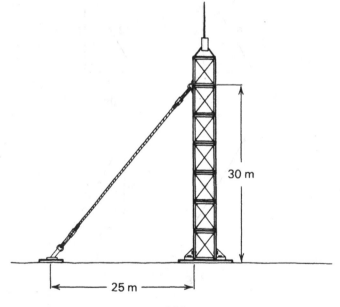

30 m

25 m

Figure 3.34

7. How far is it from second base to home plate? (*Note*: A baseball "diamond" is really a square, with sides length 90 feet.)

8. A square has area 144 sq. cm. What is the length of its diagonal?

9. One leg of a right triangle is twice the length of the other leg. The area of the triangle is 72 sq. cm. How long is the hypotenuse?

10. In the figure below, right triangle PQA_1 has two legs of length 1.
    (a) How long is hypotenuse $\overline{PA_1}$?

    Segment $\overline{A_1A_2}$ is also of length 1, drawn perpendicular to $\overline{PA_1}$.
    Segment $\overline{A_2A_3}$ is of length 1, drawn perpendicular to $\overline{PA_2}$; and so on.... .
    (b) How long is segment $\overline{PA_2}$?
    (c) How long is segment $\overline{PA_3}$?
    (d) How long is the "n-th" segment, $\overline{PA_n}$?

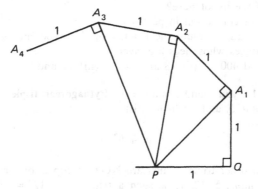

Figure 3.35

11. (a) A ship travels 6 km due South, 5 km due East, and then 4 km due South.
    How far is it from its starting point? (The answer is NOT 15 km!) You
    may assume that the path of the ship is as shown on the diagram, and
    lies in a plane.

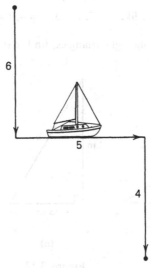

Figure 3.36

    (b) If actually the ship starts at the North Pole, on earth, what would the
    actual distance be?

12. (a) A right triangle ABC has legs of lengths 3 and 4. What is the length of the hypotenuse?

(b) Suppose we double the lengths of the legs, so that they are 6 and 8. How long is the hypotenuse in this case?

(c) Now triple the original lengths of the legs. How long is the hypotenuse in this case?

(d) Suppose we multiply the lengths of the original legs by a factor of c (c a number greater than 0), so that they have lengths $3c$ and $4c$. What is the length of the hypotenuse?

(e) Prove your conclusion in part (d).

(f) In your head, compute the lengths of the hypotenuses of the following right triangles whose legs are given:

(i) 300 and 400   (ii) 18 and 24   (iii) 27 and 36

13. In Problem 12, you found a number of **Pythagorean triples**, which are sets of three *integers* a, b, and c which satisfy

$$a^2 + b^2 = c^2.$$

Can you find other triples which are NOT multiples of the ones in Problem 12? For example, 5, 12, 13 is such a triple ($5^2 + 12^2 = 13^2$). Can you develop a calculation method which will generate Pythagorean triples? [*Hint*: Let $x = a/c$ and $y = b/c$ so $x^2 + y^2 = 1$. Use the formulas

$$x = \frac{1 - t^2}{1 + t^2} \quad \text{and} \quad u = \frac{2t}{1 + t^2}.$$

Substitute values for t, like $t = 2$, $t = 3$, $t = 4$, $t = 7$, or whatever.]

14. In each of the following right triangles, find x (think of Pythagorean triples).

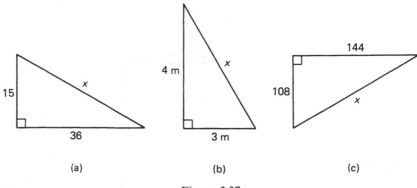

(a)   (b)   (c)

Figure 3.37

15. The lengths of the legs of a right triangle are in the ratio 2:3. If the area of the triangle is 27, how long is the hypotenuse?

16. One leg of a right triangle is 4/5 of the other. The area of the triangle is 320. Find the lengths of the legs.

17. A 20 m pole at one corner of a rectangular field is anchored by a wire stretched from the top of the pole to the opposite corner of the field as shown:

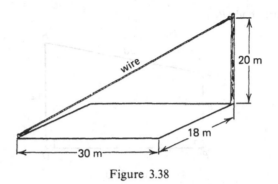

Figure 3.38

What is the length of the wire?

18. A smaller square is created inside a larger square by connecting the mid-points of the sides of the larger square, as shown:

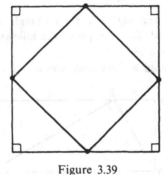

Figure 3.39

What is the ratio of the area of the small square to the area of the big square? (Is it 1/2 the big square? 1/4? $1/\sqrt{2}$? What ???) Proof?

19. A baseball diamond is a square 90 feet on a side. If a fielder caught a fly on the first base line 30 feet beyond first base, how far would he have to throw to get the ball to third base?

20. A square has area 169 cm²; what is the length of its diagonal?

21. Find the area of a square with a diagonal of length $8\sqrt{2}$.

22. Find the area of a square whose diagonal has length 8.

23. The length and width of a TV screen must be in the ratio 4:3, by FCC regu-
    lation. If a company advertises a portable TV with a screen measuring 25 cm
    along the diagonal:
    (a) What is the viewing area in square centimeters? _____
    (b) If the diagonal measurement is doubled to 50 cm, does the viewing area
        double? _____ If NO, how does it change?

24. In the figure, find $|PQ|$.

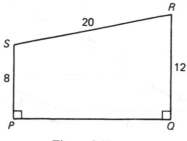

Figure 3.40

25. Use Pythagoras to prove the following statement, which is an extension of
    the **RT** postulate:

    If a leg and hypotenuse of one right triangle have the same length as a leg
    and hypotenuse respectively of another right triangle, then the third sides of
    each triangle have the same length.

26. ABCD is a rectangle, and AQD is a right triangle, as shown in the figure. If
    $|AB| = a$, $|BQ| = b$, and $|QC| = c$, prove the following:
    (a) $|AD| = \sqrt{b^2 + 2a^2 + c^2}$
    (b) $a^2 = bc$ (This should be easy once you've shown part (a).)

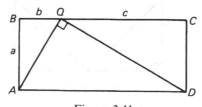

Figure 3.41

27. In triangle ABC, $|AC| = |CB|$ and $\overline{CN}$ is drawn perpendicular to $\overline{AB}$. Prove
    that $|AN| = |NB|$.

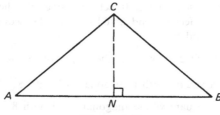

Figure 3.42

28. Let $\triangle ABC$ be a right triangle as shown on Figure 3.43, with right angle C.

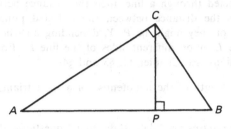

Figure 3.43

Let P be the point on $\overline{AB}$ such that $\overline{CP}$ is perpendicular to $\overline{AB}$. Prove:
(a) $|PC|^2 = |AP| \cdot |PB|$
(b) $|AC|^2 = |AP| \cdot |AB|$

29. Let $\overline{PQ}$ and $\overline{XY}$ be two parallel segments. Suppose that line L is perpendicular to these segments, and intersects the segments in their midpoint. (We shall study such lines more extensively later. The line L is called the **perpendicular bisector** of the segments.)

Given:

$L \perp \overline{PQ}$ and $L \perp \overline{XY}$

$|PM| = |MQ|$

$|XM'| = |M'Y|$

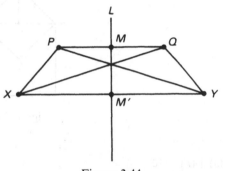

Figure 3.44

Prove:
(a) $|PX| = |QY|$
(b) $|PY| = |QX|$
[*Hint*: Use the lines perpendicular to $\overline{PQ}$ and $\overline{XY}$ passing through P and Q and intersecting $\overline{XY}$ in points P' and Q' as on Figure 3.45. Use various equalities of segments, and Pythagoras' theorem.]

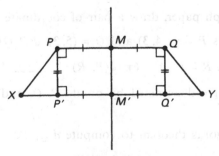

Figure 3.45

Remark. Statements (a) and (b) can be interpreted by saying that if two points are reflected through a line, then the distance between the two points is the same as the distance between the reflected points. The two points might be P, X or they might be P, Y, depending on whether they lie on one side of the line L or on different sides of the line L. For a more systematic study of reflections, see Chapter 11, §3 and §4.

30. Prove that the length of the hypotenuse of a right triangle is $\geq$ the length of a leg.

31. The next five exercises are "data sufficiency" questions. In this type of question, you are asked to make a specific calculation (in this case the area of a particular square), using the data given in the question. If the given data is sufficient to determine the answer, you do so. If it is not sufficient, you answer "insufficient data". Each question is independent of the others.

   Three squares intersect as shown in Figure 3.46. Find the area of $\triangle ADG$ for the following sets of data.

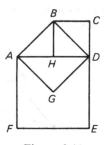

Figure 3.46

(a) $|AF| = 10$
(b) $|CE| = 18$
(c) $|BD| = 3\sqrt{2}$
(d) Area of square $BCDH = 49$
(e) Area of figure $AGDEF = 27$

EXPERIMENT 3-3

On a piece of graph paper, draw a pair of coordinate axes.

1. Draw the points $P = (-4, 3)$ and $Q = (5, 3)$. $d(P, Q) =$ _____?

2. Draw the point $R = (-4, -6)$. $d(P, R) =$ _____?

3. Draw the triangle determined by points P, Q, and R. What kind of triangle is it?

4. Use the Pythagoras theorem to compute $d(Q, R)$.

5. Plot the points $X = (-1, 7)$ and $Y = (3, -2)$. Compute $d(X, Y)$ by drawing a right triangle and using the Pythagoras theorem.

6. Let A be the point $(5, 3)$, and let X be an arbitrary point (x_1, x_2). Plot point A on the graph paper, choose a random location for point X, and label its coordinates (x_1, x_2). Try to write a formula giving $d(A, X)$ using the procedure used above in terms of x_1 and x_2. You may wish to refer to the formula giving the distance along a line stated in §2 of Chapter 2. (Theorem 2-1).

CHAPTER 4

The Distance Formula

4, §1. DISTANCE BETWEEN ARBITRARY POINTS

We shall apply the Pythagoras theorem to give us a formula for the distance between two points in terms of their coordinates. We are given a coordinate system, which we draw horizontally and vertically for convenience. The Pythagoras theorem told us:

In a right triangle, let a, b be the lengths of the legs (i.e. the sides forming the right angle), and let c be the length of the third side (i.e. the hypotenuse). Then

$$a^2 + b^2 = c^2.$$

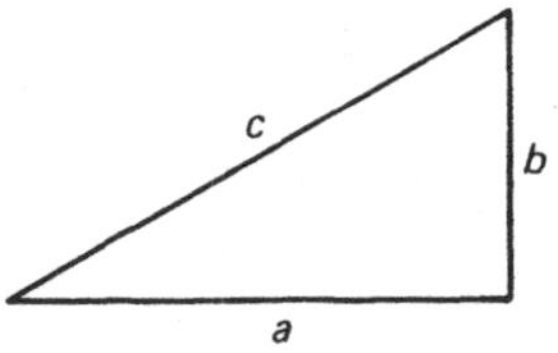

Figure 4.1

Example. Let $(1, 2)$ and $(3, 5)$ be two points in the plane. We will find the distance between them using the procedure in Experiment 3-3. First we draw the picture of the right triangle obtained from these two points, as in Figure 4.2:

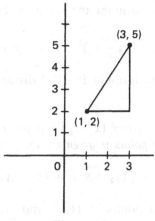

Figure 4.2

We see that the square of the length of one side is equal to

$$(3 - 1)^2 = 4.$$

The square of the length of the other side is

$$(5 - 2)^2 = (3)^2 = 9.$$

By the Pythagoras theorem, we conclude that the square of the length between the points is $4 + 9 = 13$. Hence the distance itself is $\sqrt{13}$.

Now in general, let (x_1, y_1) and (x_2, y_2) be two points in the plane. We can again make up a right triangle, as shown in Figure 4.3.

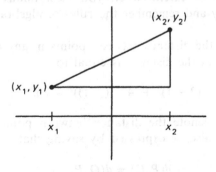

Figure 4.3

The square of the bottom side is $(x_2 - x_1)^2$, which is also equal to $(x_1 - x_2)^2$. The square of the vertical side is $(y_2 - y_1)^2$, which is also equal to $(y_1 - y_2)^2$. If d denotes the distance between the two points, then

$$d^2 = (x_2 - x_1)^2 + (y_2 - y_1)^2,$$

and therefore we get the formula for the distance between the points, namely

Theorem 4-1. *If (x_1, y_1) and (x_2, y_2) are points in the plane, then the distance d between the points is given by the formula:*

$$d = \sqrt{(x_2 - x_1)^2 + (y_2 - y_1)^2}.$$

Example. Let the two points be $(1, 2)$ and $(1, 3)$. Then the distance between them is equal to

$$\sqrt{(1 - 1)^2 + (3 - 2)^2} = 1.$$

Example. Find the distance between the points $(-1, 5)$ and $(4, -3)$. This distance is equal to

$$\sqrt{(4 - (-1))^2 + (-3 - 5)^2} = \sqrt{89}.$$

Example. Find the distance between the points $(2, 4)$ and $(1, -1)$. The square of the distance is equal to

$$(1 - 2)^2 + (-1 - 4)^2 = 26.$$

Hence the distance is equal to $\sqrt{26}$.

Warning. Always be careful when you meet minus signs. Place the parentheses correctly and remember the rules of algebra.

We can compute the distance between points in any order. In the last example, the square of the distance is equal to

$$(2 - 1)^2 + (4 - (-1))^2 = 26.$$

If we let $d(P, Q)$ denote the distance between points P and Q, then our last remark can also be expressed by saying that

$$d(P, Q) = d(Q, P).$$

Note that this is the basic property **DIST 2** of Chapter 1.

We define a point P to be **equidistant** from points X and Y if $d(P, X) = d(P, Y)$.

4, §1. EXERCISES

Find the distance between the following points P and Q. Draw these points on a sheet of graph paper.

1. $P = (2, 1)$ and $Q = (1, 5)$

2. $P = (-3, -1)$ and $Q = (-4, -6)$

3. $P = (-2, 1)$ and $Q = (3, 7)$

4. $P = (-3, -4)$ and $Q = (-2, -3)$

5. $P = (3, -2)$ and $Q = (-6, -7)$

6. $P = (-3, 2)$ and $Q = (6, 7)$

7. $P = (-3, -4)$ and $Q = (-1, -2)$

8. $P = (-1, 5)$ and $Q = (-4, -2)$

9. $P = (2, 7)$ and $Q = (-2, -7)$

10. $P = (3, 1)$ and $Q = (4, -1)$

11. Points A, B, C, D have coordinates $(0, 0)$, $(7, 0)$, $(9, 5)$, $(2, 5)$ respectively. Show that $|AB| = |DC|$ and $|AD| = |BC|$. What can you say about the quadrilateral $ABCD$?

12. A, B, C are the points $(3, 4)$, $(3, -2)$, $(-5, -2)$ respectively.
    (a) State which of the line segments joining the points is horizontal and give its length.
    (b) State which of the line segments is vertical and give its length.
    (c) Find the length of the third line segment.

13. Show that $(3, 5)$ is equidistant from $(-1, 2)$ and $(3, 0)$.

14. Find k so that the distance from $(-1, 1)$ to $(2, k)$ shall be 5 units.

15. A square has one vertex at $(4, 0)$ and its diagonals intersect at the origin. Find the coordinates of the other vertices, and find the length of the side of the square.

16. Let $K = (0, y)$ be a point on the y-axis.
    (a) If $A = (4, 3)$, write the formula for the distance from A to K.
    (b) If $B = (-8, 0)$, write the formula for the distance from B to K.
    (c) If the distance from B to K is equal to twice the distance from A to K, solve for y. (In other words, find the possible locations for point y.)

17. What are the coordinates of the point on the y-axis which is equidistant from $(-3, 2)$ and $(5, 1)$? _____

18. Prove that if $d(P, Q) = 0$, then $P = Q$. Thus we have now proved two of the basic properties of distance.

19. Let $A = (a_1, a_2)$ and $B = (b_1, b_2)$. Let r be a positive number. Write down the formula for $d(A, B)$. Define the **dilation** rA to be

$$rA = (ra_1, ra_2).$$

For instance, if $A = (-3, 5)$ and $r = 7$, then $rA = (-21, 35)$. If $B = (4, -3)$ and $r = 8$, then $rB = (32, -24)$. Prove in general that

$$d(rA, rB) = r \cdot d(A, B).$$

We shall investigate dilations more thoroughly in Chapter 8.

20. If the point $(k, 6)$ is equidistant from $(1, 1)$ and $(3, 5)$, find the value of k.

21. The vertices of a triangle are the points $(1, 8)$, $(4, 1)$ and $(7, 1)$. Find the area of the triangle.

22. The vertices of square $ABCD$ are $(4, 0)$, $(4, 4)$, $(8, 4)$ and $(8, 0)$. The area of $ABCD$ equals
    (a) 2   (b) 4   (c) 8   (d) 12   (e) 16

4, §2. HIGHER DIMENSIONAL SPACE

We have seen that a point on a line can be represented by a single number. Similarly, once a coordinate system is chosen, a point in the plane can be represented by a pair of numbers (x, y). We may do something analogous for 3-dimensional space. We choose three perpendicular axes as shown in Figure 4.4, and call them the x-axis, y-axis, and z-axis.

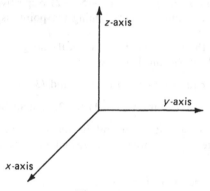

Figure 4.4

Given a point P in space, we can then represent P by a triple of numbers, $P = (x, y, z)$.

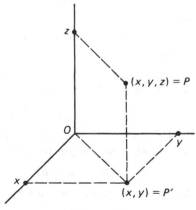

Figure 4.5

We call the numbers x, y, z the **coordinates** of the point as before.

The point $O = (0, 0, 0)$, whose coordinates are all equal to zero, is called the **origin**.

The point (x, y) with the first two coordinates only is called the **projection** of P in the (x, y)-plane.

Example. Let $P = (2, 3, 5)$. The projection of P in the (x, y)-plane is the point $(2, 3)$. Similarly, the projection of P in the (x, z)-plane is the point $(2, 5)$.

Example. Let $P = (-3, 2, -6)$. The projection of P in the (y, z)-plane is $(2, -6)$. The projection of P in the (x, z)-plane is $(-3, -6)$.

What is the distance from the origin to a point $P = (x, y, z)$? We refer to the example when we derived the diagonal of a box from its sides, in Chapter 3, §2. The situation is analogous here. Let $P' = (x, y)$ be the projection of P in the (x, y)-plane. Let t be the distance from the origin O to P'. By the Pythagoras theorem applied in the (x, y)-plane, we find

$$t^2 = x^2 + y^2.$$

On the other hand, we have a right triangle $\triangle OP'P$ whose right angle has vertex P'. Let d be the distance from O to P. Applying the Pythagoras theorem to $\triangle OP'P$ we find

$$d^2 = t^2 + z^2$$

and therefore we have the general formula

$$d^2 = x^2 + y^2 + z^2$$

giving the distance from $O = (0, 0, 0)$ to $P = (x, y, z)$.

Example. What is the distance from the origin to the point $(3, 1, 2)$? Let d be this distance. Then

$$d^2 = 3^2 + 1^2 + 2^2 = 9 + 1 + 4 = 14.$$

Hence $d = \sqrt{14}$.

Example. What is the distance from the origin to $(-2, 3, -5)$? The square of the distance is given by

$$d^2 = (-2)^2 + 3^2 + (-5)^2 = 4 + 9 + 25 = 38.$$

Hence $d = \sqrt{38}$.

Observe the negative coordinates in this example. The minus signs disappear when we square the coordinates.

Let

$$A = (a_1, a_2, a_3) \quad \text{and} \quad B = (b_1, b_2, b_3)$$

be points in 3-space. Define the **distance** between them to be

$$d(A, B) = \sqrt{(b_1 - a_1)^2 + (b_2 - a_2)^2 + (b_3 - a_3)^2}.$$

This formula is a simple extension of the distance formula from 2-space to 3-space.

Example. Let $A = (3, 1, 5)$ and $B = (1, -2, 3)$. Find the distance between A and B.

By definition, the square of the distance is

$$d(A, B)^2 = (1 - 3)^2 + (-2 - 1)^2 + (3 - 5)^2 = 4 + 9 + 4 = 17.$$

Hence $d(A, B) = \sqrt{17}$.

Note that it does not make any difference whether we compute $d(A, B)$ or $d(B, A)$, we get the same number. For instance, in the previous example,

$$d(B, A)^2 = (3 - 1)^2 + (1 - (-2))^2 + (5 - 3)^2$$
$$= 2^2 + 3^2 + 2^2$$
$$= 17.$$

Hence again, $d(B, A) = \sqrt{17}$.

The reason for this in general is that in the formula for the distance, we have

$$(b_1 - a_1)^2 = (a_1 - b_1)^2, \quad (b_2 - a_2)^2 = (a_2 - b_2)^2,$$
$$(b_3 - a_3)^2 = (a_3 - b_3)^2.$$

We were motivated to study triples of numbers like $(2, -1, 5)$ by our spatial intuition of 3-dimensional space. However, nothing prevents us from extending the concept of points and coordinates to higher dimensional space. To be systematic, we denote by $\mathbf{R}^1$ the set of all numbers. We denote by $\mathbf{R}^2$ the set of all pairs (x_1, x_2) of numbers. We denote by

$\mathbf{R}^3$ the set of all triples of numbers (x_1, x_2, x_3). We may then denote by $\mathbf{R}^4$ the set of all quadruples of numbers

$$(x_1, x_2, x_3, x_4)$$

and we call such a quadruple a point in 4-space. Of course, we cannot *draw* a point in 4-space the way we drew a point in 1-space, or in 2-space, or in 3-space. But just because we cannot draw such a point does not mean it does not exist. After all, we can write down four numbers, like

$$(1, -5, 2, 7)$$

and call that a point in 4-space. It *exists* because we have just written it down. It's that simple.

The question can arise whether it is *useful* to consider such quadruples of numbers. The answer is yes. The general principle involved here is that whenever you can assign a number to measure something, then you can call this number a **coordinate**. You get new coordinates every time you have a new way of assigning numbers. In this way, you will find that you are led not only to 4-dimensional space, but to spaces of higher dimensions.

Example. The oldest way of picking a dimension in addition to the three standard ones is time. We first select an origin for the time axis, say the birth of Christ. Then the time coordinate t is positive when associated with an event occurring after Christ, and is negative when associated with an event occurring before Christ. Suppose an airplane is flying in the year 1927 across the Atlantic, and the unit of time is the year. One could then give its four coordinates as

$$(x, y, z, 1927).$$

The idea of regarding time as a fourth coordinate is an old one. Already in the Encyclopedia of Diderot, dating back to the 18th century, d'Alembert writes in his article on "dimension":

Cette manière de considérer les quantités de plus de trois dimensions est aussi exacte que l'autre, car les lettres peuvent toujours être regardées comme représentant des nombres rationnels ou non. J'ai dit plus haut qu'il n'était pas possible de concevoir plus de trois dimensions. Un homme d'esprit de ma connaissance croit qu'on pourrait cependant regarder la durée comme une quatrième dimension, et que le produit temps par la solidité serait en quelque manière un produit de quatre dimensions; cette idée peut être contestée, mais elle a, ce me semble, quelque mérite, quand ce ne serait que celui de la nouveauté.

Encyclopédie, Vol. 4 (1754) p. 1010

Translated, this means:

> This way of considering quantities having more than three dimensions is
> just as right as the other, because algebraic letters can always be viewed as
> representing numbers, whether rational or not. I said above that it was
> not possible to conceive more than three dimensions. A clever gentleman
> with whom I am acquainted believes that nevertheless, one could view
> duration as a fourth dimension, and that the product time by solidity
> would be somehow a product of four dimensions. This idea may be chal-
> lenged, but it has, it seems to me, some merit, were it only that of being
> new.

Observe how d'Alembert refers to "a clever gentleman" when he obviously means himself. He is being rather careful in proposing what
must have been at the time a far-out idea.

That idea became more prevalent in the 20th century, as you are all
aware. Don't get the idea however that time is *the* fourth dimension.
Even the question: "Is time the fourth dimension?" is a bad one, because
it assumes implicitly that there is only one possible way of picking a
fourth dimension, and there are many ways. Whenever you can associate
a number with a situation, then this association constitutes a "dimension". Of course, in our present study, we are interested in the three dimensions of space, and especially in the two dimensions of the plane, but
there are many other ways of associating numbers with certain situations.

Example. Suppose a car is traveling on a road at 50 km/hr, at 10
o'clock some morning. We can describe this situation by the four coordinates

$$(x, y, 50, 10)$$

where (x, y) are the ordinary coordinates of the car in the plane, and the
other two coordinates $(50, 10)$ represent the additional data mentioned
above.

Example. We pick a situation from economics. Consider the indus-
tries:

$$\text{Chemical, Steel, Oil, Automobile, Farming.}$$

Take billions of dollars as units. For each industry, its profits measured
in billions of dollars per year give us a way of associating a number to
the industry. Thus, for instance,

$$(2, 4, 9, 1, 3, 1973)$$

would mean that the Chemical industry had profits of 2 billion dollars in 1973, the Steel industry had profits of 4 billion dollars in 1973, etc. In this context, negative numbers would be used to indicate losses. Thus

$$(-1, -3, 8, -2, 4, 1974)$$

would mean that the Chemical industry lost 1 billion dollars in 1974, but the Oil industry made 8 billion dollars in 1974. This data is represented by 6 coordinates, which can be viewed as a point in 6-dimensional space.

We won't go any further in these higher dimensional spaces, because in this course, we are principally concerned with the plane, in 2 dimensions, but we thought it might interest you to get a preview of notions which will appear in later years.

4, §2. EXERCISES

1. Give the value for the distance between the following:
   (a) $P = (1, 2, 4)$ and $Q = (-1, 3, -2)$
   (b) $P = (1, -2, 1)$ and $Q = (-1, 1, 1)$
   (c) $P = (-2, -1, -3)$ and $Q = (3, 2, 1)$
   (d) $P = (-4, 1, 1)$ and $Q = (1, -2, -5)$

2. Let r be a positive number, and let $A = (a_1, a_2, a_3)$. Define rA in a manner similar to the definition of Exercise 19, §1. Prove that

$$d(rA, rB) = r \cdot d(A, B).$$

4, §3. EQUATION OF A CIRCLE

Let P be a given point and r a number > 0. The **circle of radius** r **centered at** P is by definition the set of all points whose distance from P is equal to r. We can now express this condition in terms of coordinates.

Example. Let $P = (1, 4)$ and let $r = 3$. A point whose coordinates are (x, y) lies on the circle of radius 3 centered at $(1, 4)$ if and only if the distance between (x, y) and $(1, 4)$ is 3. This condition can be written as

(1) $$\sqrt{(x-1)^2 + (y-4)^2} = 3.$$

This relationship is called the equation of the circle of center $(1, 4)$ and radius 3. Note that both sides are positive. Thus this equation holds if and only if

(2) $$(x - 1)^2 + (y - 4)^2 = 9.$$

Indeed, if (1) is true, then (2) is true because we can square each side of (1) and obtain (2). On the other hand, if (2) is true and we take the square root of each side, we obtain (1), because the numbers on each side of (1) are positive. It is often convenient to leave the equation of the circle in the form (2), to avoid writing the messy square root sign. We also call (2) the equation of the circle of radius 3 centered at $(1, 4)$.

Example. The equation

$$(x - 2)^2 + (y + 5)^2 = 16$$

is the equation of a circle of radius 4 centered at $(2, -5)$. Indeed, the square of the distance between a point (x, y) and $(2, -5)$ is

$$(x - 2)^2 + (y - (-5))^2 = (x - 2)^2 + (y + 5)^2.$$

Thus a point (x, y) lies on the prescribed circle if and only if

$$(x - 2)^2 + (y + 5)^2 = 4^2 = 16.$$

Note especially the $y + 5$ in this equation.

Example. The equation

$$(x + 2)^2 + (x + 3)^2 = 7$$

is the equation of a circle of radius $\sqrt{7}$ centered at $(-2, -3)$.

Example. The equation

$$x^2 + y^2 = 1$$

is the equation of a circle of radius 1 centered at the origin. More generally, let r be a number > 0. The equation

$$x^2 + y^2 = r^2$$

is the equation of a circle of radius r centered at the origin. We can draw this circle as in Figure 4.6.

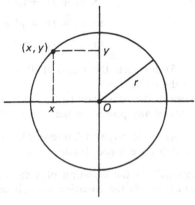

Figure 4.6

Theorem 4-2. *In general, let a, b be two numbers, and r a number > 0. Then the equation of the circle of radius r, centered at (a, b), is the equation*

$$(x - a)^2 + (y - b)^2 = r^2.$$

This means that the circle is the set of all points satisfying this equation.

Proof. We wish to find the set of all points (x, y) at distance r from (a, b). We use Theorem 4-1 to write the distance between (x, y) and (a, b) and set this distance equal to r:

$$\sqrt{(x - a)^2 + (y - b)^2} = r.$$

We then square both sides of the equation to obtain the formula above.

4, §3. EXERCISES

Write down the equation of a circle centered at the indicated point P, with radius r.

1. $P = (-3, 1)$, $r = 2$ 2. $P = (1, 5)$, $r = 3$

3. $P = (-1, -2)$, $r = 1/3$ 4. $P = (-1, 4)$, $r = 2/5$

5. $P = (3, 3)$, $r = \sqrt{2}$ 6. $P = (0, 0)$, $r = \sqrt{8}$

Give the coordinates of the center of the circle defined by the following equations, and also give the radius.

7. $(x - 1)^2 + (y - 2)^2 = 25$ 8. $(x + 7)^2 + (y - 3)^2 = 2$

9. $(x + 1)^2 + (y - 9)^2 = 8$ 10. $(x + 1)^2 + y^2 = 5/3$

11. $(x - 5)^2 + y^2 = 10$ 12. $x^2 + (y - 2)^2 = 3/2$

13. A circle with center $(3, 5)$ intersects the y-axis at $(0, 1)$.
    (a) Find the radius of the circle.
    (b) Find the coordinates of the other point of intersection with the y-axis.
    (c) State the coordinates of any points on the x-axis which are on the circle.

14. Write the equation of the circle given in Exercise 13. Give two points which lie on the circle and which have x-coordinate $= 4$.

15. Below are two "variations" on the equation of a circle. Describe the set of points in the plane which satisfy the equation in each case.
    (a) $x^2 + y^2 < 4$
    (b) $(x - 8)^2 + (y - 1)^2 = 0$

16. **Exercise in 3-space**
    The set of all points in 3-space at a given distance r from a particular point P is called the **sphere** of radius r with center P. We can write the equation for a sphere using the distance formula for 3-space given in the previous section's exercises in the same way we write the equation for a circle in 2-space.
    (a) Write down the equation for a sphere of radius 1 centered at the origin in 3-space, in terms of the coordinates (x, y, z).
    (b) Same question for a sphere of radius 3.
    (c) Same question for a sphere of radius r.
    (d) Write down the equation of a sphere centered at the given point P in 3-space, with the given radius r.
        (i) $P = (1, -3, 2)$ and $r = 1$
        (ii) $P = (1, 2, -5)$ and $r = 1$
        (iii) $P = (-1, 5, 3)$ and $r = 2$
        (iv) $P = (-2, -1, -3)$ and $r = 2$
        (v) $P = (-1, 1, 4)$ and $r = 3$
        (vi) $P = (1, 3, 1)$ and $r = 7$

17. What is the radius and the center of the sphere whose equation is
    (a) $x^2 + (y - 1)^2 + (z + 2)^2 = 16$
    (b) $(x + 3)^2 + (y - 2)^2 + z^2 = 7$
    (c) $(x + 5)^2 + y^2 + z^2 = 2$
    (d) $(x - 2)^2 + y^2 + (z + 1)^2 = 8$

CHAPTER 5

Some Applications of Right Triangles

In this chapter, we will recognize the strength and utility of the Pythagoras theorem. It is surely one of the basic "building blocks" of much mathematics.

EXPERIMENT 5-1

Recall that a point X is **equidistant** from points A and B if $d(X, A) = d(X, B)$; in other words, if its distance from A is the same as its distance from B.

1. Locate a number of points which are equidistant from points A and B on your paper (there are more than one). Can you describe carefully where these points are in relation to segment $\overline{AB}$?

2. Draw another segment $\overline{AB}$ which is 10 cm long. Find the point X on $\overline{AB}$ which divides the segment exactly in half, and draw a line through this point which is perpendicular to $\overline{AB}$ (use your protractor or the method given in Construction 1-4).

   Now choose an arbitrary point M on this line, and measure $d(M, A)$ and $d(M, B)$. Measure $\angle AMX$ and $\angle BMX$. Repeat with another point M. What can you conclude? Start thinking about how you might prove your conclusions.

3. Here's another problem to start thinking about. You will be able to answer it readily after reading the next section.

   Two factories A and B are located along a straight river bank. They wish to build a single loading dock on the bank, sharing its use and expense. At what point on the bank should the dock be located so that the distance from the dock to factory A is the same as the distance from the dock to factory B. (See Figure 5.1)

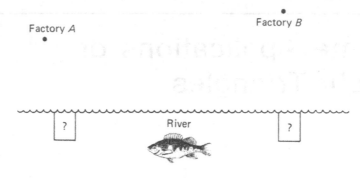

Figure 5.1

5, §1. PERPENDICULAR BISECTOR

In Experiment 5-1 you were working with the **perpendicular bisector of a segment**, which is defined to be the line perpendicular to the segment passing through the segment's midpoint. The precise definition of the **midpoint** O of a line segment $\overline{AB}$ is that it is the point on $\overline{AB}$ such that $d(A, O) = d(O, B)$. We also say that the midpoint of a segment **bisects** the segment.

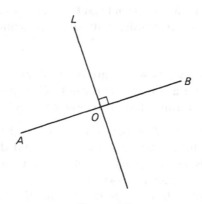

Figure 5.2

Figure 5.2 illustrates a segment $\overline{AB}$, its midpoint O, and perpendicular bisector line L.

We next want to prove an important property of perpendicular bisectors which you may have already discovered for yourself in the experiment. In order to do this, we will need to use a new property about points and line segments, and so we prove this result first. Such a "mini-theorem" which necessarily precedes and aids the proof of a regular theorem is called a **lemma**, a word borrowed from the ancient Greek mathematicians.

Lemma. *Given two distinct points P and Q. Let X be a point on line L_{PQ} such that $d(P, X) = d(Q, X)$. Then X lies on segment $\overline{PQ}$.*

Proof. Suppose X is *not* on $\overline{PQ}$. Then there are two possible situations, as illustrated in Figure 5.3.

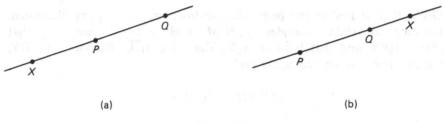

(a)   (b)

Figure 5.3

Looking at the situation in Figure 5.3(a), **SEG** tells us that

$$d(X, P) + d(P, Q) = d(X, Q).$$

Now we are given that $d(X, P) = d(X, Q)$, so we can cancel, and we are left with:

$$d(P, Q) = 0.$$

But this is impossible, since we were given that P and Q were distinct points.

Similarly, it can be proved that the situation in Figure 5.3(b) is also impossible, and this is left as an exercise at the end of this section.

Therefore, X must lie on segment $\overline{PQ}$.

If you have just been turned off because we wasted a page in this book by proving the "obvious", you may exercise your taste, make it a postulate, and forget the proof.

We can now prove the desired theorem about perpendicular bisectors. To make the language simpler we shall say (as we did in the experiment) that a point X is **equidistant** from two points P and Q if $d(P, X) = d(X, Q)$.

Theorem 5-1. *Let P, Q be distinct points in the plane. Let M also be a point in the plane. We have*

$$d(P, M) = d(Q, M)$$

if and only if M lies on the perpendicular bisector of $\overline{PQ}$.

Proof. Again, we have two "if-then" statements to consider. We will first prove the statement:

If M is on the perpendicular bisector of segment $\overline{PQ}$, *then M is equidistant from points P and Q.*

Given that M lies on the perpendicular bisector L of $\overline{PQ}$, as illustrated. Looking at right triangles $\triangle POM$ and $\triangle QOM$, we see that $|PO| = |QO|$, and that both triangles share leg $\overline{MO}$. Therefore, by **RT**, the two hypotenuses are equal, and

$$d(P, M) = d(Q, M)$$

making M equidistant from P and Q.

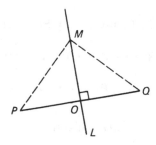

Figure 5.4

We now prove the other "if-then" statement (the converse):

If M is a point equidistant from points P and Q, then M lies on the perpendicular bisector of $\overline{PQ}$.

We are given a point M and points P and Q with

$$d(P, M) = d(Q, M).$$

As far as we know, point M may be almost anywhere in relation to P and Q; we certainly cannot *assume* that it will lie on the perpendicular bisector of $\overline{PQ}$. And so, in Figure 5.5 we have drawn a couple of "exaggerated" pictures illustrating possible situations.

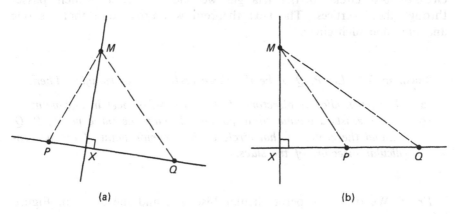

(a)   (b)

Figure 5.5

In each case, we have drawn the line through M perpendicular to L_{PQ}, intersecting L_{PQ} at X. We want to show that this perpendicular is in fact the perpendicular bisector.

By Pythagoras, we have:

$$|PX|^2 + |XM|^2 = |PM|^2$$
$$= |QM|^2 \quad \text{since } d(P, M) = d(Q, M)$$
$$= |XM|^2 + |QX|^2 \quad \text{by Pythagoras again.}$$

Canceling like terms in this final equation:

$$|PX|^2 + \cancel{|XM|^2} = \cancel{|XM|^2} + |QX|^2$$

we see that:

$$|PX|^2 = |QX|^2$$

and therefore that:

$$|PX| = |QX|$$

since we are dealing with distances, which cannot be negative.

We now have that X is a point on L_{PQ}, where

$$d(P, X) = d(Q, X).$$

By our lemma, X must lie on segment $\overline{PQ}$ (eliminating a situation like that in Figure 5.5), and therefore it is the midpoint of $\overline{PQ}$, making the perpendicular through M the perpendicular bisector.

This proves our theorem.

We shall now apply perpendicular bisectors in a situation having to do with a triangle. Let P, Q, A be the three vertices of a triangle. By a **circumscribed** circle to the triangle we mean a circle which passes through these vertices. The next theorem will prove that there is one and only one such circle.

Theorem 5-2. *Let P, Q, A be the three vertices of a triangle. Then:*

(a) *The perpendicular bisectors of the three sides meet in one point.*

(b) *There exists a unique circle passing through the three points P, Q, A, and the center of that circle is the common point of the perpendicular bisectors of the sides.*

Proof. We draw the perpendicular bisectors and the circle in Figure 5.6.

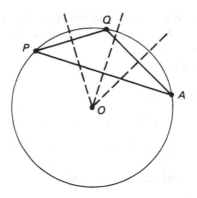

Figure 5.6

Let us now prove (a). Let O be the point of intersection of the perpendicular bisectors of $\overline{PQ}$ and of $\overline{QA}$. By Theorem 5-1 we have

$$d(O, P) = d(O, Q) \quad \text{and} \quad d(O, Q) = d(O, A).$$

Hence $d(O, P) = d(O, A)$. Again by Theorem 5-1, this implies that O is on the perpendicular bisector of $\overline{PA}$, and proves (a).

Let $r = d(O, P) = d(O, Q) = d(O, A)$. Then the circle centered at O having radius r passes through the three points P, Q, and A, so there is one circle passing through the three points. Finally, suppose that C is any circle passing through the three points, and let M be the center of C. Then

$$d(M, P) = d(M, Q) = d(M, A).$$

By Theorem 5-1 it follows that M lies on the perpendicular bisectors of $\overline{PQ}$, $\overline{QA}$, and $\overline{PA}$, so we must have $M = 0$. This proves the theorem.

5, §1. EXERCISES

1. In Figure 5.7, line L is the perpendicular bisector of segment $\overline{PQ}$. Find the lengths a, b, and c.

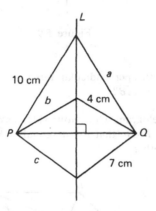

Figure 5.7

2. Line K is the perpendicular bisector of $\overline{PQ}$, and line L is the perpendicular bisector of $\overline{QM}$ in Figure 5.8. Prove that $d(0, P) = d(0, M)$.

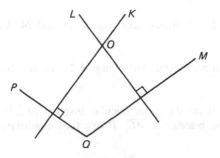

Figure 5.8

3. In Figure 5.9

$$|AX| = |AY|,$$

$$L \perp \overline{XY}.$$

Is L the perpendicular bisector of XY? Give your reasons.

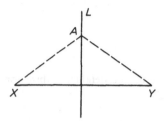

Figure 5.9

4. Given that $\overline{AB}$ is on the perpendicular bisector of $\overline{XY}$. Is $\overline{XY}$ then on the perpendicular bisector of $\overline{AB}$?

5. A **rhombus** is a parallelogram whose four sides have the same length, as illustrated in Figure 5.10. Prove that the diagonals $\overline{PM}$ and $\overline{QN}$ are perpendicular bisectors of each other.

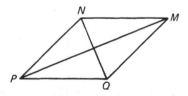

Figure 5.10

6. The diagonals of a rhombus have lengths 18 and 24. Find the lengths of the sides.

7. The diagonals of a rhombus have lengths 12 cm and 16 cm. Find the lengths of the sides.

8. Given that $\overline{AC}$ is on the perpendicular bisector of $\overline{BD}$, and that $\overline{BD}$ lies on the perpendicular bisector of $\overline{AC}$. Prove that quadrilateral $ABCD$ is a rhombus.

9. In Figure 5.11, $d(X, A) = d(Y, A)$ and $d(X, B) = d(Y, B)$. Show that L_{AB} is the perpendicular bisector of $\overline{XY}$.

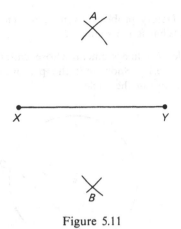

Figure 5.11

10. KIDS has gone into the television business, and they now broadcast a TV signal to an area of radius 100 km.

YGLEPH•

KIDS transmitter •

ZYZZX •

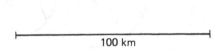

100 km

Figure 5.12

A cable TV operator wants to serve Ygleph and Zyzzx by cable. By law, he must locate his receiving tower as far way from the KIDS transmitter as possible (and still receive the signal). He also wants to locate the tower so that the length of cable from the tower to Ygleph is the same as the length of cable from the tower to Zyzzx. How should he determine the tower's location?

11. Given three points P, Q, and R in the plane, not on the same line. Describe how you would find a point X which is equidistant from all three points. Is there more than one location for X?

12. Prove the second part of the lemma.

13. Look back at the factory problem given at the end of the experiment begin-ning this section. What is the solution?

14. A **chord** of a circle is a line segment whose endpoints are on the circle, as illustrated in Figure 5.13. Show that the perpendicular bisector of a chord goes through the center of the circle.

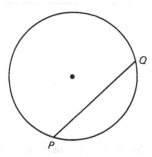

Figure 5.13

15. Show that the line through the center of a circle perpendicular to a chord bisects the chord.

16. A circle centered at P and a circle centered at Q intersect at two points, X and Y, as shown in Figure 5.14. Prove that segment $\overline{PQ}$ goes through the midpoint of segment $\overline{XY}$.

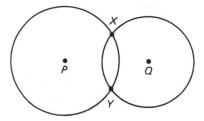

Figure 5.14

CONSTRUCTION 5-1

Constructing the perpendicular bisector of a line segment (or bisecting a line segment).

We wish to construct the perpendicular bisector of a given line seg-ment $\overline{AB}$.

On your paper, draw an arbitrary line segment $\overline{AB}$. Set your compass tips at a convenient distance which is longer than half the length of the segment. With the point of the compass at A, draw an arc above the segment. Repeat with the compass point at *B*, keeping the same setting, producing a figure as below. Label the point where the arcs intersect *X*.

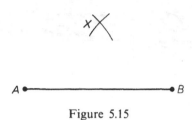

A •————————————————————• B

Figure 5.15

Now using the same compass setting (or a different one if you wish) repeat the procedure given above but draw the arcs below line segment $\overline{AB}$, labeling the point of intersection *Y*. (If you use a different compass setting, you could also put the second pair of arcs above the line segment.)

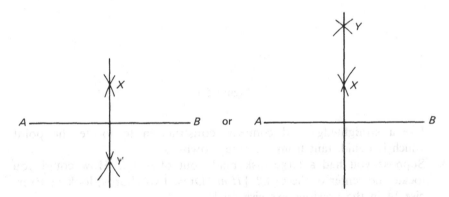

Figure 5.16

The line L_{XY} is the perpendicular bisector. Look back at the set of exercises preceding this construction; which one proves that this line is indeed the perpendicular bisector? Since the perpendicular bisector of a segment passes through the midpoint of the segment, we also bisected line segment $\overline{AB}$.

1. Draw a large triangle on your paper using a straightedge. Construct a circle which passes through the three vertices of the triangle. [*Hint*: You need to locate a point which is at the same distance from each of the vertices. This distance is the radius of the circle. Use perpendicular bisectors to locate this point.]
2. As chief engineer for Dusty Desert Power and Light Co., you've been asked to determine the location of a new distributing station which

must be equidistant from the towns of Zyzzx, Ygleph, and Pfeifle. A map is shown below:

The Dusty Desert

Figure 5.17

Use a straightedge and compass construction to locate the point which is equidistant from the three towns.

3. Suppose you had a large disk made out of metal. How could you locate the center of the disk? [*Hint*: Draw two chords; look at Exercise 14 in the previous exercise set.]

CONSTRUCTION 5-2

The perpendicular line through a given point to a given line.

We wish to construct the line perpendicular to a given line L, passing through a point P *not on L*:

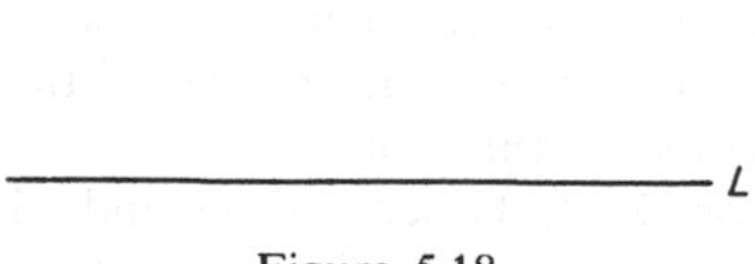

•P

————————————— L

Figure 5.18

Set your compass tips so that, with the point of the compass on P, you can draw two arcs which intersect line L, as shown:

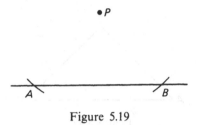

Figure 5.19

Label these intersection points A and B. Using this same setting (or a different one) place the point of the compass on A and draw an arc below segment $\overline{AB}$. Repeat with the compass point at B. Label the intersection point of these two arcs Q.

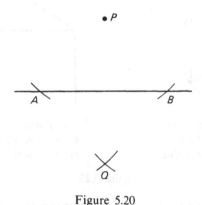

Figure 5.20

1. The line through P and Q is the desired perpendicular. Can you prove why?

2. Construct a line through point P which is parallel to line L. [Hint: Use the theorem that says that two lines perpendicular to the same line are parallel.]

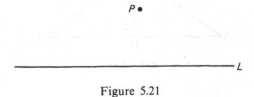

Figure 5.21

5, §2. ISOSCELES AND EQUILATERAL TRIANGLES

In Figure 5.22 we illustrate a triangle ABC where $|AB| = |BC|$.

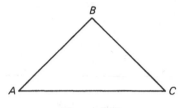

Figure 5.22

Such a triangle, where two sides have the same length, is called an **isosceles** triangle. Isosceles triangles seem to crop up often in mathematics as well as in practical situations; two examples are illustrated below:

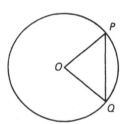

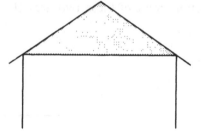

In the circle centered at O, $\triangle POQ$ is isosceles since $|OP| = |OQ| = r$ the radius.

The portion of the wall beneath most pitched roofs is an isosceles triangle region.

Figure 5.23

In any isosceles triangle, the point common to the two equal length sides (point B in Figure 5.24) lies on the perpendicular bisector of the third side, since it is equidistant from the endpoints of the third side. Therefore, the height of the triangle drawn from this "equidistant" point to the third side is the perpendicular bisector of the third side, and thus divides it into two equal parts:

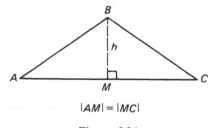

$|AM| = |MC|$

Figure 5.24

This is a handy situation, as it makes calculation of the area of an isosceles triangle possible knowing only the lengths of the sides.

Example. Find the area of triangle XYZ, whose sides have lengths as indicated.

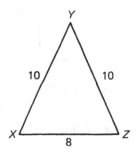

Figure 5.25

We draw the height from vertex Y, meeting side $\overline{XZ}$ at M.

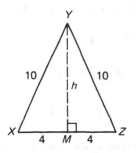

Figure 5.26

Since $|XM| = |MZ|$, we have $|XM| = 4$. By Pythagoras,

$$|XM|^2 + h^2 = |XY|^2,$$

and so

$$4^2 + h^2 = 10^2,$$

$$h = \sqrt{84}.$$

The area of the triangle then is $\frac{1}{2} \cdot 8 \cdot \sqrt{84} = 4\sqrt{84}$ square units.

Example. The front of a house looks like a square together with an isosceles triangle under the roof, as shown in Figure 5.27, with the indicated dimensions.

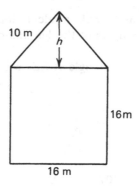

Figure 5.27

What is the area of the front of the house?

The area of the square is $16 \cdot 16 = 256 \text{ m}^2$.

Let h be the height of the roof as shown. By Pythagoras, we know that

$$8^2 + h^2 = 10^2 = 100.$$

Hence

$$h^2 = 100 - 64 = 36.$$

Therefore $h = 6$. Hence the area of the triangle under the roof is equal to

$$\tfrac{1}{2} \cdot 16 \cdot 6 = 48.$$

Adding the area of the square and the area of the triangle yields the total area, which is

$$256 + 48 = 304 \text{ m}^2.$$

Such a computation is useful when you want to paint the house. You know how much paint is needed for each square meter, and you can then compute how much paint you will need to do the house.

Example. We know enough now to *prove* that Construction 1-3 duplicates an angle as stated. In fact we are reduced to proving the following statement:

Let $\triangle POQ$ and $\triangle P'O'Q'$ be two isosceles triangles whose corresponding sides have the same measure, that is suppose that:

$$|PO| = |OQ| = |P'O'| = |O'Q'| \qquad and \qquad |PQ| = |P'Q'|.$$

Then

$$m(\angle POQ) = m(\angle P'O'Q').$$

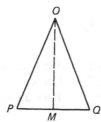 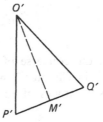

Figure 5.28

Proof. Let $\overline{OM}$ be the perpendicular bisector of $\overline{PQ}$, intersecting $\overline{PQ}$ in M. Similarly, let $\overline{O'M'}$ be the perpendicular bisector of $\overline{P'Q'}$, intersecting $\overline{P'Q'}$ in M'. Then

$$|PM| = \tfrac{1}{2}|PQ|$$
$$= \tfrac{1}{2}|P'Q'| \quad \text{by assumption}$$
$$= |P'M'|.$$

The right triangles $\triangle PMO$ and $\triangle P'M'O'$ have hypotenuses of equal length by assumption, and sides $\overline{PM}$ and $\overline{P'M'}$ of equal length by what we just proved. By Pythagoras, it follows that $|OM| = |O'M'|$. By **RT**, it follows that

$$m(\angle POM) = m(\angle P'O'M').$$

Similarly $m(\angle MOQ) = m(\angle M'O'Q')$. Hence $m(\angle POQ) = m(\angle P'O'Q')$, thereby proving what we wanted.

Theorem 5-3 (Isosceles Triangle Theorem). *Given an isosceles triangle PQR with $|PQ| = |QR|$. Then the angles opposite to the sides $\overline{PQ}$ and $\overline{QR}$ have the same measure. In other words: $m(\angle P) = m(\angle R)$.*

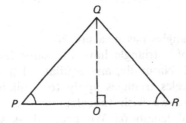

Figure 5.29

Proof. By Theorem 5-1, the height $\overline{QO}$ of the triangle bisects $\overline{PR}$, making $|PO| = |OR|$. By **RT**, applied to $\triangle POR$ and $\triangle QOR$, we conclude that

$$m(\angle P) = m(\angle R).$$

Incidentally, the converse of this theorem is also true, and we will be able to prove it in Chapter 7, §2, Theorem 7-3. For convenience, we call the equal-measure angles of an isosceles triangle **base angles**.

Example. In isosceles triangle $\triangle ABC$, $m(\angle B) = 40°$. How large are base angles $\angle A$ and $\angle C$?

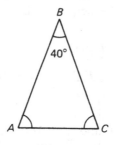

Figure 5.30

We know from the previous chapter that:

$$m(\angle A) + m(\angle C) + 40° = 180°.$$

Since $m(\angle A) = m(\angle C)$, we have

$$2 \cdot m(\angle A) + 40° = 180°,$$
$$m(\angle A) \qquad = 70°.$$

So each of the base angles has measure 70°.

If all three sides of a triangle have the same length, we say that the triangle is **equilateral**. Naturally, any equilateral triangle is also isosceles, so properties of isosceles triangles apply to equilateral triangles as well. For example, suppose we wish to calculate the area of an equilateral triangle with sides of length 6. We proceed as we did with isosceles triangles:

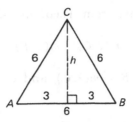

Figure 5.31

We use Pythagoras to calculate height h:

$$h^2 + 3^2 = 6^2,$$

$$h^2 = 27,$$

$$h = 3\sqrt{3}.$$

Then

$$\text{area of } ABC = \tfrac{1}{2}6h$$

$$= \tfrac{1}{2} \cdot 6 \cdot 3\sqrt{3}$$

$$= 9\sqrt{3}.$$

Also, we can prove a result about the angles of an equilateral triangle using the previous theorem about isosceles triangles. A result like the next one, which can be deduced almost immediately from a previous theorem, is called a **corollary** (from the Latin word for "gift").

Corollary to Theorem 5-3. *All the angles of an equilateral triangle have the same measure, which is* 60°.

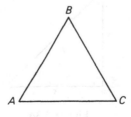

Figure 5.32

Proof. We have equilateral triangle $\triangle ABC$, with $|AB| = |BC| = |CA|$. Since $|AB| = |BC|$, we have from Theorem 5-3 that

$$m(\angle A) = m(\angle C).$$

Since $|CA| = |AB|$ we have from Theorem 5-3 again (turn your head) that:

$$m(\angle C) = m(\angle B).$$

Therefore, $m(\angle A) = m(\angle B) = m(\angle C)$, and each one must be one-third of 180°, which is 60°.

Again, we will be able to prove the converse of this statement when more powerful techniques are at hand.

In trigonometry and construction, a triangle which is useful as a "reference" is the **isosceles right triangle**, where both legs along the right angle have the same length, as illustrated:

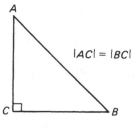

Figure 5.33

By our isosceles triangle theorem, we can see that

$$m(\angle A) = m(\angle B) = 45°.$$

Suppose we have an isosceles right triangle with legs of length s, as illustrated below:

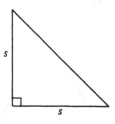

Figure 5.34

If we let d be the length of the hypotenuse, by Pythagoras we have:

$$d^2 = s^2 + s^2$$
$$= 2s^2,$$

therefore

$$d = \sqrt{2s^2}$$
$$= \sqrt{2} \cdot s.$$

This relationship is a useful one.

> *In an isosceles right triangle with legs of*
> *length s, the hypotenuse has length* $\sqrt{2} \cdot s$.

Example. What is the length of the diagonal of a square with side length 12?

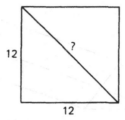

Figure 5.35

The diagonal is the hypotenuse of an isosceles right triangle with legs of length 12. Thus its length is $12\sqrt{2}$.

5, §2. EXERCISES

1. Triangle ABC is isosceles, with $|AB| = |AC|$. For each of the values of $m(\angle A)$ below, find $m(\angle B)$ and $m(\angle C)$:
   (a) $m(\angle A) = 40°$ (b) $m(\angle A) = 60°$ (c) $m(\angle A) = 90°$
   (d) $m(\angle A) = 110°$.

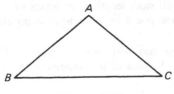

Figure 5.36

2. Find the height to base $\overline{BC}$ and the area of isosceles triangle ABC if its dimensions are as follows:
   (a) $a = 5$; $b = 3$ (b) $a = 10$; $b = 8$ (c) $a = 5$; $b = 2$

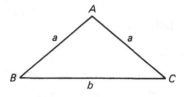

Figure 5.37

3. A cabin has dimensions as indicated below. Compute the area of the surface formed by the sides and roof. Don't forget the triangular pieces just below the roof. The bottom of the cabin touching the ground is not to be counted.

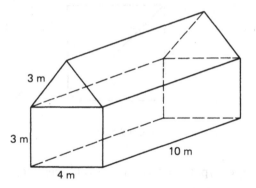

Figure 5.38

4. What is the area of an equilateral triangle whose sides have length:
   (a) 6 cm   (b) 3 m   (c) $\sqrt{3}$ m

5. If an equilateral triangle has sides of length s, find a formula which will give you the height of the triangle in terms of s. [*Hint*: Draw a picture, then use Pythagoras.]

6. Looking at your result from Exercise 5, write a formula giving the area of an equilateral triangle with sides length s, in terms of s. This is a good formula to know, as it will save you a lot of time in computing these areas.

7. An equilateral triangle has perimeter 12 meters. Find its area. (The **perimeter** of a triangle is the sum of the lengths of its sides.)

8. An equilateral triangle has perimeter 24 meters. Find its area.

9. The far wall of your bedroom has dimensions as shown in Figure 5.39. You want to paint it in metal-flake pink paint, which comes in spray cans costing $4.50 each. On the label, it says that each can will cover 9 sq. m of surface. How much are you going to spend on this project?

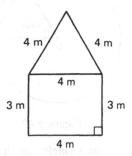

Figure 5.39

10. Find x in each of the triangles below:

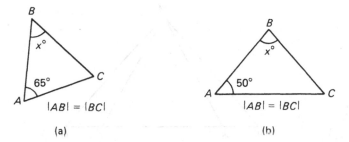

(a)

(b)

Figure 5.40

11. In Figure 5.41, $m(\angle MOP) = 90°$, $|MO| = |OP| = 1$, and $|MP| = |PQ|$.

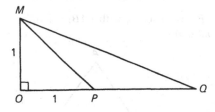

Figure 5.41

(a) What is the length of $\overline{MQ}$? _____ .

(b) $m(\angle Q) =$ _____ .

(c) $m(\angle QMO) =$ _____ .

12. In the figure, the circle centered at O has radius 8. If $m(\angle AOB) = 60°$, find the area of triangle AOB. [*Hint*: What kind of a triangle is AOB?]

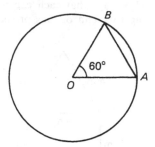

Figure 5.42

13. Show that if an isosceles right triangle has hypotenuse length x, then the length of either leg is $x/\sqrt{2}$.

14. Let $\triangle PMQ$ be an isosceles triangle, with $|PM| = |MQ|$. Let X and Y be distinct points on $\overline{PQ}$ such that $d(P, X) = d(Y, Q)$. Prove that $\triangle XMY$ is also isosceles.

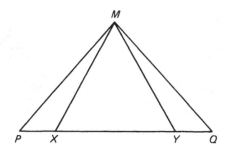

Figure 5.43

15. When you double the lengths of the sides of an equilateral triangle, the area increases by a factor of _____?
(DON'T GUESS. Use the formula you derived in Exercise 6.)

16. Given: Triangle ABC is isosceles, with $|AB| = |AC|$.
Prove: $m(\angle x) = m(\angle y)$.

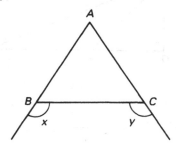

Figure 5.44

17. In Figure 5.45, $|BC| = |CA| = |AD|$. Prove that $m(\angle EAD) = 3 \cdot m(\angle B)$.

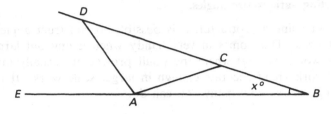

Figure 5.45

18. Triangle QED is isosceles, with $|QE| = |ED|$. Segment $\overline{IT}$ is drawn parallel to base $\overline{QD}$. Prove that $m(\angle x) = m(\angle y)$.

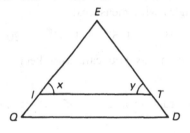

Figure 5.46

19. Prove that the length of the hypotenuse of a right triangle is greater than the length of a leg. [*Hint*: Draw an arbitrary right triangle; write down the related Pythagoras equation. Suppose that the hypotenuse *was* the same length as a leg, then derive a contradiction. Similar procedure to show that the hypotenuse cannot be shorter than a leg.]

20. In an isosceles triangle ABC where $|AB| = |BC|$, prove that the line through B, perpendicular to $\overline{AC}$, bisects $\angle ABC$; in other words, prove that $m(\angle ABO) = m(\angle CBO)$.

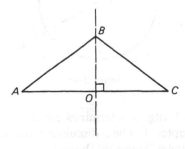

Figure 5.47

21. Use your result from Exercise 20 to prove that the method of bisecting an angle given in Construction 1-2 works. [*Hint*: Re-do the construction. Draw segment $\overline{PQ}$.]

CONSTRUCTION 5-3

Constructing various size angles.

Without using a protractor, it is possible to construct angles of many different sizes. This comes in very handy when laying out large projects (such as woodworking) where the small protractors usually intended for detailed work aren't accurate enough in larger scale work. It is also just a challenge to see how many you can do.

Look over the results in the previous section, and determine methods that you could use to construct angles with the following measures:

(a) 45°   (b) 60°   (c) 30°   (d) 15°

By constructing two angles adjacent to one another, you should be able to construct angles with measures:

75°   105°   135°   120°

and others. Do as many as you can, and keep a record of how to do them!

Example. To construct an angle of 45°, bisect an angle of 90°.

5, §3. THEOREMS ABOUT CIRCLES

Given a circle with center O, and two points P and Q on the circle. As usual, rays R_{OP} and R_{OQ} determine two angles, and these angles are called **central angles** (since the vertex O of each is the center of a circle).

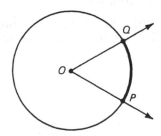

Figure 5.48

Each of the central angles determines an arc on the circle, in a manner discussed in Chapter 1 when discussing the naming of angles. We will say the central angle **intercepts** the arc.

It will be convenient in this context to say that an **arc has x degrees** to mean that the central angle which intercepts the arc has x degrees. In the example below, we say arc $\overset{\frown}{PQ}$ has 65° since central angle POQ has

65°: Thus we define the **measure of an arc** to be the measure of its central angle.

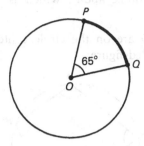

Figure 5.49

Let P, Q, and M be three points on a circle. The angle lying between the rays R_{MQ} and R_{MP} is called the **inscribed angle** $\angle PMQ$. We illustrate some inscribed angles.

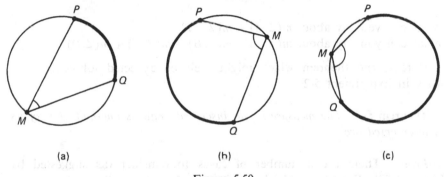

(a)   (h)   (c)

Figure 5.50

Note that an inscribed angle intercepts an arc on the circle, as indicated above. In the next experiment, you will discover a relationship between the measures of an inscribed angle and its arc.

EXPERIMENT 5-2

This experiment leads to Theorem 5-4.

Draw a number of large circles with your compass, and in each draw an arbitrary inscribed angle. Try to get some variation, including each of the situations pictured in Figure 5.50.

In each example, measure the inscribed angle with a protractor, and write down the number of degrees in each. Now determine the number of degrees in the arc intercepted by each inscribed angle. Do this by drawing the central angle which intercepts the same arc and measure it with a protractor.

What can you conclude about the measure of an inscribed angle and the measure of its intercepted arc?

Here's a problem to think about, which is related to the previous exercise:

Points A, B, C, and D are on the circle centered at O, and are connected to make a four-sided figure:

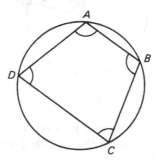

Figure 5.51

What can you say about $m(\angle A) + m(\angle C)$?

What can you say about $m(\angle A) + m(\angle B) + m(\angle C) + m(\angle D)$?

Here is the theorem which might well be expected following your work in Experiment 5-2.

Theorem 5-4. *The measure of an inscribed angle is one-half that of its intercepted arc.*

Proof. There are a number of cases to consider (as suggested by Figure 5.50). We let $\angle PMQ$ be an inscribed angle in all cases.

Case I. One ray of the inscribed angle passes through the center of the circle. We suppose this ray is R_{MQ} as shown on Figure 5.52(a), (b).

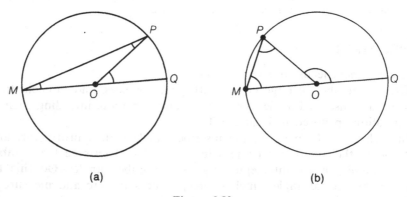

(a)   (b)

Figure 5.52

Since the sum of the measures of the angles of a triangle has 180°, we know that

$$m(\angle M) + m(\angle P) + m(\angle MOP) = 180°,$$

and we also have by the hypothesis of Case I

$$m(\angle POQ) + m(\angle MOP) = 180°.$$

Hence by algebra,

$$m(\angle M) + m(\angle P) = m(\angle POQ).$$

But $|OM| = |OP|$ (why?), so by the Isosceles Triangle Theorem we conclude that

$$m(\angle M) = m(\angle P).$$

Hence

$$2m(\angle M) = m(\angle POQ), \qquad \text{so} \qquad m(\angle M) = \tfrac{1}{2}m(\angle POQ),$$

thus proving Case I.

   Case II. The points P and Q lie on opposite sides of the line L_{MO}.

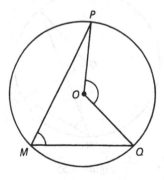

Figure 5.53

We wish to show that $m(\angle M) = \tfrac{1}{2}m(\angle POQ)$.

For the proof, draw L_{MO} which divides the inscribed angle M and central angle $\angle POQ$ into two parts and intersects the circle in point R:

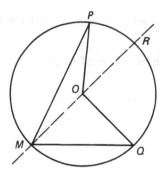

Figure 5.54

Then we get:

$$m(\angle PMQ) = m(\angle PMO) + m(\angle OMQ)$$
$$= \tfrac{1}{2}m(\angle POR) + \tfrac{1}{2}m(\angle ROQ) \quad \text{by Case I}$$
$$= \tfrac{1}{2}[m(\angle POR) + m(\angle ROQ)]$$
$$= \tfrac{1}{2}m(\angle POQ).$$

This proves the theorem in Case II.

Case III. The points P and Q lie on the same side of the line L_{MO}.

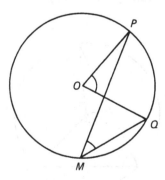

Figure 5.55

Again we want to prove that $m(\angle M) = \tfrac{1}{2}m(\angle POQ)$.

Once again we draw line L_{MO} intersecting the circle in point R.

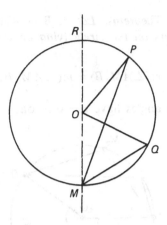

Figure 5.56

Then we use the same method as in Case II, but with a subtraction instead of an addition, namely:

$$m(\angle PMQ) = m(\angle OMQ) - m(\angle OMP)$$

$$= \tfrac{1}{2}m(\angle ROQ) - \tfrac{1}{2}m(\angle ROP) \quad \text{by Case I}$$

$$= \tfrac{1}{2}[m(\angle ROQ) - m(\angle ROP)]$$

$$= \tfrac{1}{2}m(\angle POQ),$$

thus proving Case III.

This concludes the proof of our theorem. If you are wondering what happened to the situation illustrated in Figure 5.50(c), notice that it is just an exaggerated version of Case II.

Example. In Figure 5.57(a) the measure of each angle $\angle M_1$, $\angle M_2$, $\angle M_3$ is $\tfrac{1}{2} \cdot 90$, since they all determine the same arc as central angle $\angle O$. So this measure is 45°.

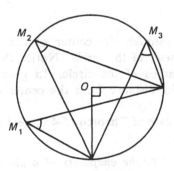

Figure 5.57(a)

Example (Apollonius Theorem). *Let A, B be distinct points on a circle. Let M_1, M_2 be points on the circle lying on the same arc between A and B. Then*

$$m(\angle AM_1B) = m(\angle AM_2B).$$

This is because both angles have measure one-half of $m(\angle AOB)$.

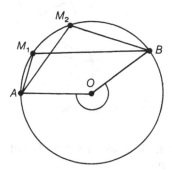

Figure 5.57(b)

A **chord** is a line segment whose endpoints are on the circle. Figure 5.58 illustrates some chords:

 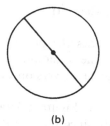

(a)   (b)

Figure 5.58

A chord which passes through the center of the circle (as in Figure 5.58(b)) is called a **diameter** of the circle. Notice that the length of any diameter is twice the radius of the circle. In general, the length of a chord is dependent on its distance from the center of the circle, as you will see in the exercises.

An important special case of Theorem 5-4 is next stated as a theorem.

Theorem 5-5. *Let P, Q be the endpoints of a diameter of a circle. Let M be any other point on the circle. Then $\angle PMQ$ is a right angle.*

Proof. Work this out as a special case of Theorem 5-4; in other words, do Exercise 3.

If we have a circle centered at a point O and we are given a point P on the circle, we define the **tangent to the circle** at P to be the line through P perpendicular to the radius drawn from O to P.

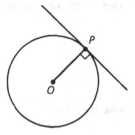

Figure 5.59

It is an important property of this tangent that it intersects the circle only at the point P. You should prove this as Exercise 8, referring back to Theorem 3-5.

Warning. Although it has been repeated many times in the last few thousand years that the tangent to a curve at a point is the line which touches the curve only at that point, this statement is generally false. It is true only in special cases like the circle. Examples when it is false are given in the next figure.

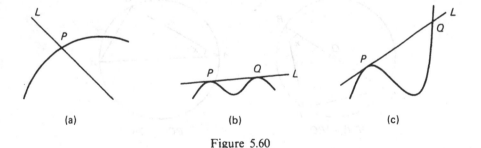

Figure 5.60

In Figure 5.60(a), the line L intersects the curve in only one point P, but is not tangent. In Figure 5.60(b), line L intersects the curve in two points P, Q, and is tangent. In Figure 5.60(c), line L intersects the curve in two points P, Q. It is tangent at P and not tangent at Q. Also, don't you want to say that a line is tangent to itself? But a line touches itself in all of its points! So it certainly touches itself in more than one point.

These examples show the extent to which stupidities can be repeated un-critically over thousands of years. You can learn how to handle tangents to more general curves in more advanced courses like calculus.

Also of interest are the formulas for the area of a disk and the cir-cumference of a circle (the distance "around" the circle). We delay the derivations of these formulas until Chapter 8, when we can use the theory of dilations to make our work very easy.

5, §3. EXERCISES

1. Find x in each of the following diagrams:

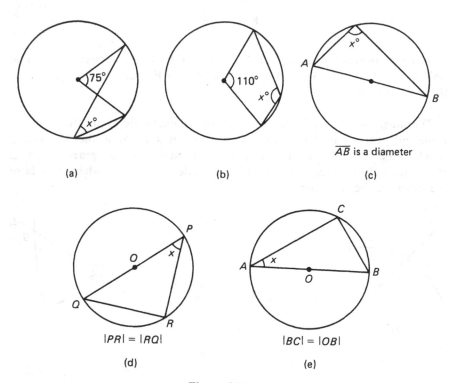

Figure 5.61

2. Referring to Figure 5.62, what is the sum of the measures of the arcs $\overset{\frown}{XY}$, $\overset{\frown}{YZ}$, $\overset{\frown}{ZW}$, and $\overset{\frown}{WX}$?

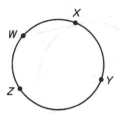

Figure 5.62

3. In Figure 5.63, $\overline{PQ}$ is a diameter, M is any other point on the circle. In tri-angle $\triangle PMQ$, prove that $\angle M$ is a right angle.

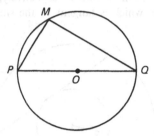

Figure 5.63

4. In Figure 5.64, the circle centered at O has radius 10 cm. If the distance from O to chord $\overline{AB}$ is 8 cm, how long is chord $\overline{AB}$? (Recall that the dis-tance from a point to a line is defined to be the length of the segment from the point perpendicular to the line.)

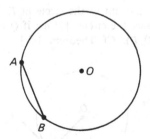

Figure 5.64

5. Same question as in Exercise 4, except that the distance from O to the chord is 3 cm.

6. In Figure 5.65, $d(O, P) = 15$ and line L is tangent to the circle. How long is $\overline{PQ}$ if the radius of the circle is 12?

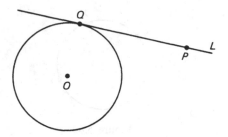

Figure 5.65

7. We say that two circles are **concentric** if they have the same center. Two concentric circles have radii 3 and 7 respectively, as shown. What is the length of the chord $\overline{AB}$, which is tangent to the smaller circle?

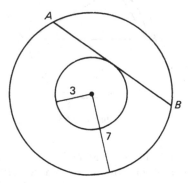

Figure 5.66

8. In Figure 5.67, line L is tangent to the circle at P. Prove that no other point on L can possibly intersect the circle. [*Hint*: If Q is a point on L and $P \neq Q$, show that $d(O, Q) > d(O, P)$. Cf. Theorem 3-5.]

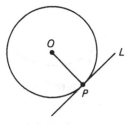

Figure 5.67

9. Suppose that two lines tangent to a circle at points P and Q intersect at a point M, as shown. Prove that $|PM| = |QM|$.

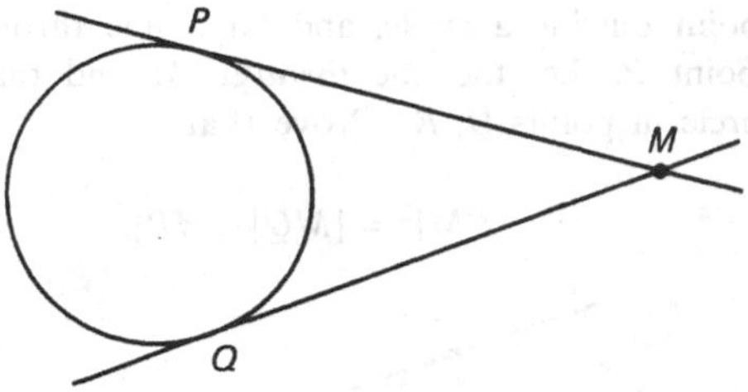

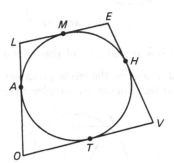

Figure 5.68

10. The sides of figure *LOVE* are tangent to the circle, as shown:

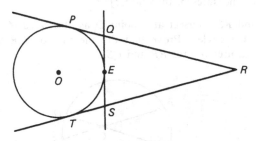

Figure 5.69

Prove that $|OV| + |LE| = |LO| + |VE|$.

11. In Figure 5.70, all three lines are tangent to the circle. If $|PR| = 5$ cm, what is the perimeter of $\triangle QRS$?

Figure 5.70

[*Hint*: Let E be the point where $\overline{QS}$ is tangent to the circle. How do $|PQ|$ and $|QE|$ differ?]

12. Let M be a point outside a circle, and let a line through M be tangent to the circle at point P. Let the line through M and the center of the circle intersect the circle in points Q, R. Prove that

$$|PM|^2 = |MQ| \cdot |MR|.$$

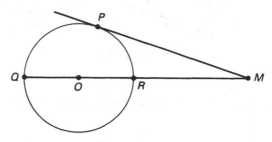

Figure 5.71

(Actually, Exercise 9 is a special case of this relationship.)

13. Points P, U, N, and T are on the circle given below. Prove that the opposite angles of the four-sided figure are supplementary.

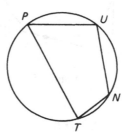

Figure 5.72

[*Hint*: Look at the arcs intercepted by the opposite inscribed angles. What is the sum of the measures of these arcs?]

14. Chords PQ and RS intersect at a point O as illustrated (O is not necessarily the center of the circle). Prove that the measure of $\angle ROP$ is one-half the sum of the measures of arcs RP and QS.

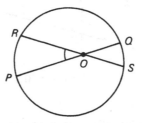

Figure 5.73

[*Hint*: Draw $\overline{PS}$. Look at inscribed angles $\angle RSP$ and $\angle QPS$, and triangle *POS*.]

15. Two rays are drawn from the same point A outside a circle, and intersect the circle as shown in the picture at right. Prove that the measure of $\angle A$ is one-half the difference between the measures of arcs $\overset{\frown}{BD}$ and $\overset{\frown}{CE}$.

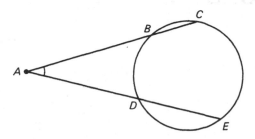

Figure 5.74

[*Hint*: Draw $\overline{DC}$. Look at inscribed angles $\angle BCD$ and $\angle CDE$, and triangle *ACD*.]

CHAPTER 6

Polygons

6, §1. BASIC IDEAS

Figure 6.1 illustrates some polygons:

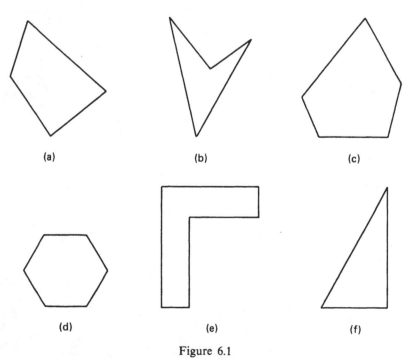

(a)

(b)

(c)

(d)

(e)

(f)

Figure 6.1

Figure 6.2 illustrates some figures which are NOT polygons:

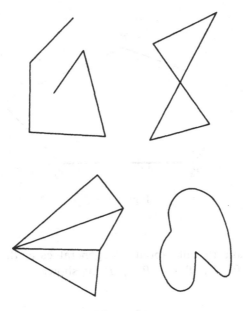

Figure 6.2

We shall give the definition of a polygon in a moment. In these figures, observe that a polygon consists of line segments which enclose a single region.

A four-sided polygon is called a **quadrilateral** (Figure 6.1(a) or (b)). A five-sided polygon is called a **pentagon** (Figure 6.1(c)), and a six-sided polygon is called a **hexagon** (Figure 6.1(d)). If we kept using special prefixes such as quad-, penta-, hexa-, and so on for naming polygons, we would have a hard time talking about figures with many sides without getting very confused. Instead, we call a polygon which has n sides an **n-gon**. For example, a pentagon could also be called a 5-gon; a hexagon would be called a 6-gon. If we don't want to specify the number of sides, we simply use the word polygon (poly- means many).

As we mentioned for triangles (3-gons), there is no good word to use for the region inside a polygon, except "polygonal region", which is a mouthful. So we shall speak of the area of a polygon when we mean the area of the polygonal region, as we did for triangles.

If a segment $\overline{PQ}$ is the side of a polygon, then we call point P or point Q a **vertex** of the polygon. With multisided polygons, we often label the vertices (plural of vertex) P_1, P_2, P_3, etc. for a number of reasons. First, we would run out of letters if the polygon had more than 26 sides. Second, this notation reminds us of the number of sides of the polygon; in the illustration, we see immediately that the figure has 5 sides:

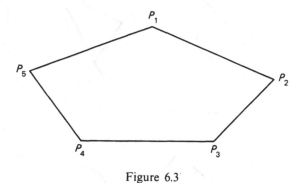

Figure 6.3

Finally, if we want to talk about the general case, the n-gon, we can label its vertices P_1, P_2, P_3, ...,P_{n-1}, P_n as shown:

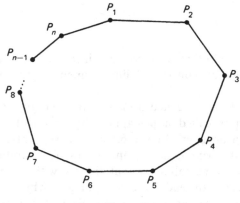

Figure 6.4

We can now define a **polygon** (or an **n-gon**) to be an n-sided figure consisting of n segments

$$\overline{P_1P_2}, \quad \overline{P_2P_3}, \quad \overline{P_3P_4}, ...,\overline{P_{n-1}P_n}, \quad \overline{P_nP_1},$$

which intersect only at their endpoints and enclose a single region.

EXPERIMENT 6-1

Below are two rows of polygons. Each polygon in the top row exhibits a common property, while those in the bottom row do not.

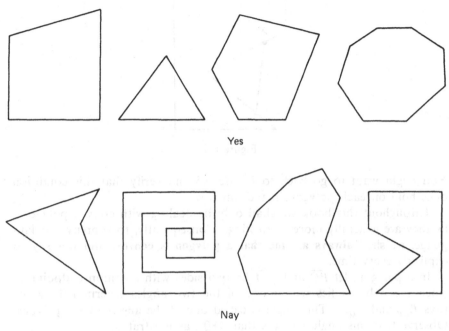

Yes

Nay

Figure 6.5

Can you discover what the top row polygons have in common that the bottom ones do not? Try to state the definition of your property as specifically as possible, using terms and concepts that we have already defined. There are many possible answers.

6, §2. CONVEXITY AND ANGLES

Polygons which look like those in the top row of Figure 6.5 we will call convex. Thus we define a polygon to be **convex** if it has the following property:

Given two points X and Y on the sides of the polygon, then the segment $\overline{XY}$ is wholly contained in the polygonal region surrounded by the polygon (including the polygon itself).

Observe how this condition fails in a polygon such as one chosen from the lower row in Figure 6.5:

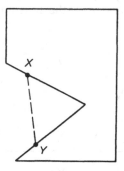

Figure 6.6

You might want to go back to Figure 6.5 and verify that this condition does hold on each polygon in the top row.

Throughout this book we shall only be dealing with convex polygons, as they are generally more interesting. Consequently, to simplify our language, **we shall always assume that a polygon is convex**, and not say so explicitly every time.

In a polygon, let $\overline{PQ}$ and $\overline{QM}$ be two sides with common endpoint Q. Then the polygon lies within one of the two angles determined by the rays R_{QP} and R_{QM}. This angle is called one of the **angles of the polygon**. Observe that this angle has less than 180°, as illustrated:

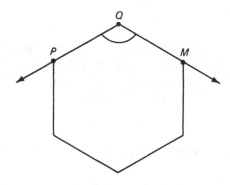

Figure 6.7

EXPERIMENT 6-2

1. Besides the number of sides, two characteristics of polygons are the lengths of its sides and the measures of its angles.
   (a) What do we call a quadrilateral which has four sides of the same length and which has four angles with the same measure?

(b) Can you think of a quadrilateral which has four angles with equal measures but whose sides do not all have the same length? Draw a picture. What do we call such a quadrilateral?

(c) Can you draw a quadrilateral which has four sides of equal length, but whose angles do not have the same measure?

(d) What do we call a 3-gon which has equal length sides and equal measure angles?

2. With a ruler, draw an arbitrary looking convex quadrilateral. Measure each of its four angles, and add these measures. Repeat with two or three other quadrilaterals.

3. Repeat the procedure given in Part 2 with a few pentagons, and then a few hexagons.

What can you conclude? Can you say what the sum of the measures of the angles of a 7-gon would be? How about a 13-gon?

For the rest of this experiment, we will develop a formula to answer these questions.

Consider a convex quadrilateral. A line segment between two opposite vertices is called a **diagonal**. We can decompose the quadrilateral ("break it down") into two triangles by drawing a diagonal, as shown:

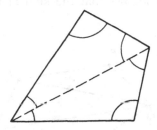

Figure 6.8

Notice that the angles of the two triangles make up the angles of the polygon. What is the sum of the angles in each triangle? In the two triangles added together? In the polygon?

Now look at a convex pentagon. We can decompose it into triangles, using the "diagonals" from a single vertex, as shown:

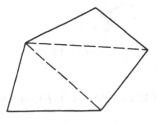

Figure 6.9

We see that in a 5-gon we get three such triangles. Again, the angles of the triangles make up the angles of the polygon when it is decomposed in this way. What is the sum of the measures of all the angles in the triangles? What is the sum of the measures of all the angles in the polygon?

Repeat this procedure with a hexagon to find the sum of the measures of its angle. Continue the process until you can state a formula which will give the sum of the measures of the angles of an *n*-gon in terms of *n*. If you have succeeded, you will have found the next theorem.

Theorem 6-1. *The sum of the angles of a polygon with n sides has*

$$(n - 2)180°.$$

Proof. Let $P_1, P_2, \ldots, P_n$ be the vertices of the polygon as shown in the figure. The segments

$$\overline{P_1 P_3}, \ \overline{P_1 P_4}, \ \ldots, \overline{P_1 P_{n-1}}$$

decompose the polygon into $n - 2$ triangles. Since the sum of the angles of a triangle has 180°, it follows that the sum of the angles of the polygon has $(n - 2)180°$.

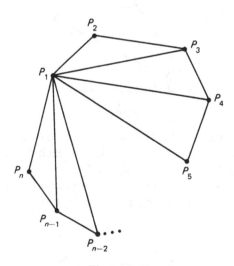

Figure 6.10

Example. The sum of the angles of a 7-gon has

$$(7 - 2)180° = 5 \cdot 180° = 900°.$$

By definition, the segments $\overline{P_1P_3}$, ..., $\overline{P_1P_{n-1}}$ are called **diagonals** of the polygon.

6, §3. REGULAR POLYGONS

A polygon is called **regular** if all its sides have the same length and all its angles have the same measure. For example, a square is also a regular 4-gon. An equilateral triangle is a regular 3-gon.

The **perimeter** of a polygon is defined to be the sum of the lengths of its sides.

Example. The perimeter of a square whose sides have length 9 cm is equal to $9 \cdot 4 = 36$ cm.

Example. Find the perimeter of a regular 11-gon, with sides of length 5 cm. The perimeter $= 11 \cdot 5 = 55$ cm.

Example. If each angle of a regular polygon has 135°, how many sides does the polygon have?

Let n be the number of sides. This is also the number of vertices. Since the angle at each vertex has 135°, the sum of these angles has $135n°$. By Theorem 6-1, we must have

$$135n = (n - 2)180.$$

By algebra, this is equivalent with

$$135n = 180n - 360,$$

and we can solve for n. We get

$$360 = 180n - 135n = 45n,$$

whence

$$n = \frac{360}{45} = 8.$$

The answer is that the regular polygon has 8 sides.

A polygon whose vertices lie on a circle is said to be **inscribed** in the circle. We illustrate a regular hexagon inscribed in a circle below:

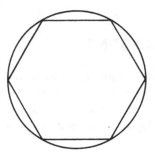

Figure 6.11

It is an interesting problem to try to construct various *n*-gons by inscribing them in circles using only a compass and straightedge. To construct a regular hexagon, square, or octagon is fairly easy. A regular pentagon is difficult. Some *n*-gons are impossible!

CONSTRUCTION 6-1

Suppose we wish to construct a regular hexagon. First observe that our regular hexagon above can be decomposed into six equilateral triangles:

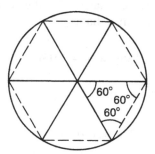

Figure 6.12

This means that the length of a side of the hexagon is the same as the radius of the circle. Knowing this, we pick a point P_1 on the circle. Set the compass at a distance equal to the radius of the circle. Put the tip on P_1. Draw an arc intersecting the circle in a point which you label P_2. Set the tip on P_2. Draw an arc intersecting the circle in a new point P_3. Continue in this manner until you have found all the six vertices of the hexagon.

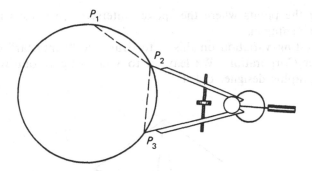

Figure 6.13

To construct a square, draw a diameter, and then construct another diameter perpendicular to the first (refer to Construction 1-4 of Chapter 1). Connecting the endpoints of the diameters will produce the square.

We leave it to you to construct a regular octagon, and any other *n*-gon you think you can figure out.

If we allow ourselves the use of a protractor, there is an easy way to draw any regular *n*-gon in a circle. The procedure is to draw radii like spokes of a wheel at intervals of $360°/n$. We then connect the points where the radii intersect the circle.

Example. If we wish to draw a regular pentagon, we divide the full angle by the number of sides:

$$\frac{360°}{5} = 72°.$$

We then draw "spokes" emanating from the center of our circle at intervals of 72°. Use a protractor.

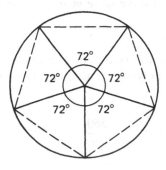

Figure 6.14

Connecting the points where the spokes intersect the circle will produce our regular pentagon.

An interesting variation on this is to draw the "pentastar" emblem of the Chrysler Corporation. We leave it to you to figure out exactly how Chrysler's graphic designer did it.

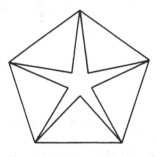

Figure 6.15

When you have finished drawing various polygons using the protractor, use your technique to create a design similar to the pentastar based on regular polygons.

Inscribed Polygons

In Construction 6-1 you constructed regular *n*-gons by inscribing them in a circle. We give here a proof of the general fact behind this construction.

Theorem 6-2. *Given an inscribed polygon in a circle, with vertices $P_1, P_2, \ldots, P_n$. Assume that the central angles formed with any two successive vertices all have the same measure. Then the polygon is a regular polygon.*

Proof. The radii from the center of the circle to the vertices will be called the spokes of the polygon. The measure of the angle between the spokes will then be

$$\frac{360°}{n}.$$

We illustrate this general case below.

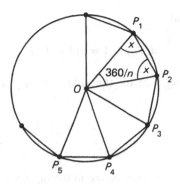

Figure 6.16

We need to show that the measures of the angles of the polygons are equal and that the lengths of the sides are equal.

Let's first deal with the angles. Each one of the triangles $\triangle P_1 O P_2$, $\triangle P_2 O P_3$, and so on is an isosceles triangle whose sides have lengths equal to the radius of the circle. If x denotes the measure of the base angles of one of these isosceles triangles, then we must have:

$$2x + 360/n = 180°,$$

$$2x = 180 - 360/n,$$

$$x = 90 - 180/n.$$

This measure is the same for all the triangles. Therefore each angle of the constructed polygon has a measure equal to $2x$, which is:

$$180 - 360/n.$$

Thus for a pentagon, each angle has measure:

$$180 - 360/5 = 180 - 72 = 108°.$$

To see that the sides of the constructed polygon have the same length, observe that O lies on the perpendicular bisector of the segment $\overline{P_1 P_3}$, because $d(O, P_1) = d(O, P_3)$. Furthermore, in an earlier exercise you proved that the perpendicular bisector is also the bisector of the angle $P_1 O P_3$. Since P_2 lies on this angle bisector, we can conclude that P_2 lies on the perpendicular bisector, and that

$$d(P_1, P_2) = d(P_2, P_3).$$

Similarly, we prove that

$$d(P_2, P_3) = d(P_3, P_4)$$

and so on, thus showing that all the sides of our polygon have the same length.

6, §3. EXERCISES

1. Determine whether or not each of the following figures is a polygon. If it is, state whether or not it is convex:

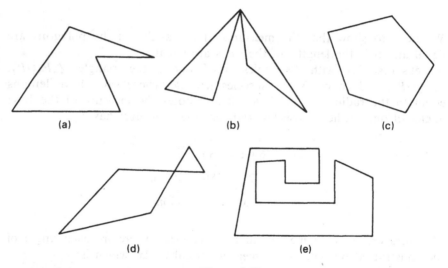

Figure 6.17

2. What is the sum of the measures of the angles of:
   (a) an octagon   (b) a pentagon   (c) a 12-gon

3. Find the measure of each angle of a regular polygon with:
   (a) 6 sides   (b) 11 sides   (c) 14 sides   (d) n sides

4. Is it possible to have a regular polygon each angle of which has 153°? Give reasons for your answer.

5. If each angle of a regular polygon has 165°, how many sides does the polygon have?

6. How many sides does a polygon have if the sum of the measures of the angles is:
   (a) 2700°   (b) 1080°   (c) $d°$

7. If each angle of a regular polygon has 140°, how many sides does the polygon have?

8. Give an example of a polygon whose sides all have the same length, but which is not a regular polygon.
   Give an example of a polygon whose angles all have the same measure, but which is not a regular polygon.

9. An isosceles triangle has a side of length 10, and a side of length 4. What is its perimeter? (Yes, there *is* enough information given.)

10. The sides of a triangle have lengths $2n - 1$, $n + 5$, $3n - 8$ units.
    (a) Find a value of n for which the triangle is isosceles.
    (b) How many values of n are there which make the triangle isosceles?
    (c) Is there a value of n which makes the triangle equilateral?

11. An equilateral triangle has perimeter 36. Find its area.

12. A square has perimeter 36. Find its area.

13. In the figure, angle X is measured on the outside of an *arbitrary* n-gon. Prove that the sum of all n such outside angles equals $(n + 2)180$. [*Hint*: What is the sum of all n interior angles?]

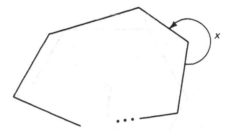

Figure 6.18

14. Squares of side length x are cut out of the corners of a 4 cm × 5 cm piece of sheet metal, as illustrated:

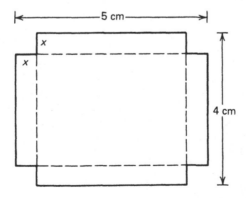

Figure 6.19

Show that the perimeter of the piece of metal stays constant, regardless of the value of x, as long as $x < 2$.

15. What is the area of the piece of metal in Exercise 14?

16. Prove that the area of a regular hexagon with side of length s is:

$$\frac{3s^2\sqrt{3}}{2}.$$

[*Hint*: Divide the hexagon into triangles by drawing "radii". What kind of triangles are they?]

17. What is the area of a regular hexagon whose perimeter is 30 cm?

18. The length of each side of a regular hexagon is 2. Find the area of the hexagon.

19. In the figure, $ABCDEF$ is a regular hexagon. The distance from the center O to any side is x; length of each side is s. Prove that the area of the hexagon is $3xs$.

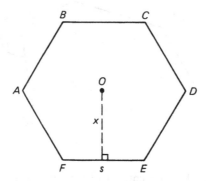

Figure 6.20

CHAPTER 7

Congruent Triangles

In previous chapters we have considered triangles which have special features like isosceles triangles or equilateral triangles. Now we come to studying arbitrary triangles.

7, §1. EUCLID'S TESTS FOR CONGRUENCE

First, what do we mean by "congruence"? Look at the two triangles in Figure 7.1. They have the same dimensions.

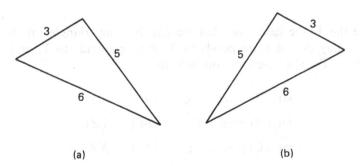

Figure 7.1

In what other respects are they the same triangles? They are not "equal". For instance, the triangle in (a) points to the left while the triangle in (b) points to the right. Do you think this difference is important? For the type of questions we wish to consider, it turns out not to be important. We might be tempted to say that they are "equal". This would be a bad use of the word. We have already mentioned that in mathematics, when we say that two things are equal we mean that they are *the same*. On the other hand, the above triangles are alike in many ways: the lengths of their sides are equal; the measures of their angles are equal; their perimeters are equal; their areas are equal. We might say that one is an "exact copy" of the other.

In this chapter we shall use the informal definition: Two figures are called **congruent** if we can lay one exactly on the other without changing its shape. Then the triangles of Figure 7.1 are congruent. In this chapter we shall apply the notion of congruence only to triangles. In Chapters 11 and 12 you will see how to deal with other figures.

The symbol for congruence is $\simeq$. Thus to denote that two triangles $\triangle ABC$ and $\triangle XYZ$ are congruent, we write

$$\triangle ABC \simeq \triangle XYZ.$$

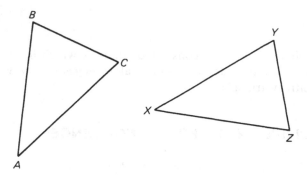

Figure 7.2

Suppose this is the case, and that we can lay one triangle on the other in such a way that A corresponds to X, B corresponds to Y, and C corresponds to Z. Then we can conclude that:

$$m(\angle A) = m(\angle X), \qquad |AB| = |XY|,$$
$$m(\angle B) = m(\angle Y), \qquad |BC| = |YZ|,$$
$$m(\angle C) = m(\angle Z), \qquad |AC| = |XZ|.$$

Knowing that two triangles are congruent therefore gives us at least six bits of information, namely that the corresponding sides and angles have the same measure.

In addition, we can often "decompose" more complex figures into triangles. If we can establish that certain of these triangles are congruent, we can discover relationships in the more complex figures. We apply this idea to parallelograms, for example.

The main question is, what is the minimum amount of information we need to know about two triangles to be able to conclude that they are congruent? Experiment 7-1 will give us some clues.

EXPERIMENT 7-1

Use ruler, compass, and protractor to do the following.

1. Construct a triangle with sides of lengths 4 cm, 6 cm, 8 cm. Can you construct another triangle with sides of lengths 4 cm, 6 cm, 8 cm which is not congruent to the first one? If yes, do so.

2. Construct a triangle which has a side of length 6 cm, a side of length 8 cm, and an angle with 30°. Can you construct another triangle with these dimensions which is not congruent? If yes, do so. Any others?

3. Suppose that we require the 30° angle in Step 2 to be placed between the sides having lengths 6 cm and 8 cm. How many non-congruent triangles are there which satisfy this additional requirement?

4. Construct a triangle that has a 45° angle and a 60° angle. How many different sized or shaped triangles can you construct with these requirements?

5. Construct a triangle which has a 45° angle, a 60° angle, and a side of length 6 cm. How many different sized or shaped triangles can you construct with these dimensions?

6. If we require that the 6 cm side lie between the 45° angle and the 60° angle in Step 5, how many different shaped triangles could be constructed with this additional requirement?

7. How many different right triangles are there with sides of length 2 cm and 3 cm ("sides" also includes the hypotenuse)?

We can now attack the question of how much information we must have concerning two triangles in order to conclude that they are congruent.

Suppose we know that two triangles each have sides of lengths 5 cm, 6 cm, and 7 cm; we illustrate one in Figure 7.3.

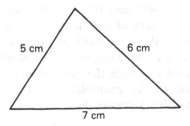

Figure 7.3

In Question 1 of the experiment, you probably realized that there is only one possible shape for a triangle given the lengths of its sides. Thus any other triangle with the same dimensions as the triangle in Figure 7.3 would have to be congruent. We take this property as postulate for this chapter:

SSS. *If the three sides of a triangle have the same lengths as the three corresponding sides of another triangle, then the triangles are congruent.*

(SSS stands for **side-side-side.)**

  Example. Given two triangles △PQR and △XYZ with dimensions as shown.

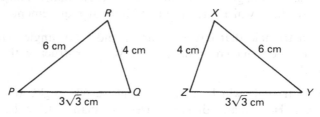

Figure 7.4

We can conclude that △PQR ≃ △XYZ by **SSS**. It is clear that P corresponds to Y, Q corresponds to Z, and R corresponds to X. Thus we may also conclude that

$$m(\angle P) = m(\angle Y), \qquad m(\angle Q) = m(\angle Z), \qquad m(\angle R) = m(\angle X).$$

These pairs of angles are the corresponding angles.

  Suppose we know that two triangles each have a side of length 3 cm and angles with measures 45° and 60°. What may we conclude in this situation? First, since the sum of measures of the angles of any triangle is 180°, we can conclude that the third angle in each triangle has 75°. But unless we know where the 3 cm side is in relation to the angles in both triangles, we cannot conclude that the triangles are congruent. The two triangles in Figure 7.5, for example, meet the requirements but are not congruent (you may have drawn these triangles in Question 5 of the experiment).

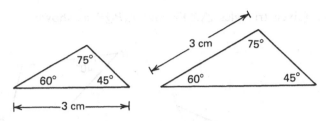

Figure 7.5

On the other hand, suppose we require that the equal length side be located in each triangle between the corresponding angles. Then it turns out that the triangles are congruent (Figure 7.6).

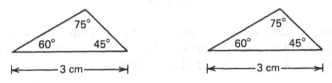

Figure 7.6

This is a concrete case of Euclid's second criterion for congruence.

ASA. *If two triangles have one corresponding side of the same length, and two corresponding angles of the same measure, then the triangles are congruent.*

(**ASA** stands for **angle-side-angle**.)

Example. Given triangles △*PQR* and △*ABC* as shown:

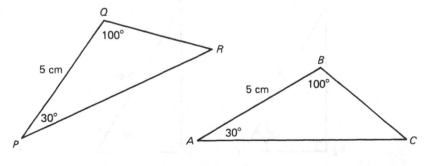

Figure 7.7

By ASA, we can conclude that △*PQR* ≃ △*ABC*. This then tells us that $|PR| = |AC|$ and $|QR| = |BC|$.

Example. Given triangles $\triangle XYZ$ and $\triangle PQR$ as shown:

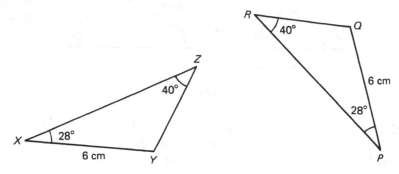

Figure 7.8

We first notice that $m(\angle Y) = m(\angle Q) = 112°$. Thus $\overline{XY}$ and $\overline{PQ}$ are corresponding sides, and we apply **ASA**. Since $\triangle XYZ \simeq \triangle PQR$, we know also that $|XZ| = |PR|$ and $|YZ| = |RQ|$.

Remark. In **ASA** we did not assume that the corresponding side is the common side to the two angles. However, if we had formulated **ASA** by making this assumption we could deduce the stronger version. Indeed, if two angles of a triangle are known, then the third angle is also known because the sum of the measures of the angles of a triangle is 180°.

Example. Suppose we know two right triangles have hypotenuses with the same length and corresponding acute angles with the same measure, as shown:

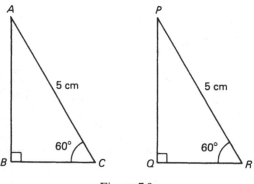

Figure 7.9

By an argument like that in the previous example, we can conclude that $\triangle ABC \simeq \triangle PQR$. Thus we see that one pair of equal length corresponding sides and one pair of equal measure acute angles in a *right* triangle is sufficient for congruence.

Having considered the situation where we have two angles and one side with the same measure in two triangles, we now consider the case of two sides and one angle having the same measure. Look back at Questions 2 and 3 in the experiment. In order to insure that the triangles meeting these requirements are congruent it seems that we have to require, in addition, that the angles lie between the two given sides. With this additional requirement, the triangles are congruent.

SAS. *If two sides and the angle between them in one triangle have the same measures as two sides and the angle between them in another triangle, then the triangles are congruent.*

(**SAS** stands for **side-angle-side**.)

Example. In the figure, with the given dimensions, we have

$$\triangle HIM \simeq \triangle XYZ.$$

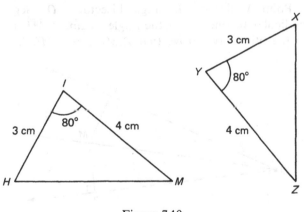

Figure 7.10

A common exercise in classical geometry is to prove that a triangle in a given figure is congruent to some other triangle in the figure. The hardest part of these proofs is trying to decide which of the postulates about congruent triangles (**SSS**, **SAS**, or **ASA**) is most applicable.

Example. Line segments $\overline{PQ}$ and $\overline{RS}$ intersect at O and bisect each other as shown below. Prove that

$$\triangle POR \simeq \triangle QOS.$$

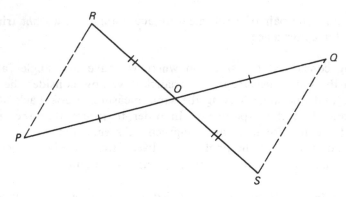

Figure 7.11

Proof. By Theorem 1-1 we know that the vertical angles $\angle QOS$ and $\angle ROP$ have the same measure. By hypothesis, we know that

$$|RO| = |OS| \quad \text{and} \quad |PO| = |OQ|.$$

We can then use **SAS** to conclude that $\triangle POR$ is congruent to $\triangle QOS$.

Example. Point M lies on the angle bisector $\angle O$. Segment $\overline{MQ}$ is drawn perpendicular to one ray of the angle; segment $\overline{MP}$ is drawn perpendicular to the other ray. Prove that $\triangle MOQ \simeq \triangle MOP$.

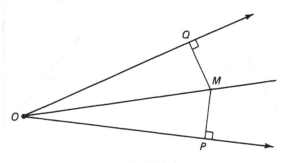

Figure 7.12

Proof. Since ray R_{OM} bisects $\angle O$, we have

$$m(\angle QOM) = m(\angle POM).$$

Both triangles share side $\overline{OM}$, which we notate:

$$|OM| = |OM|.$$

Since $\angle OQM$ and $\angle OPM$ are right angles, we have

$$m(\angle OQM) = m(\angle OPM).$$

At this point, we mark on the figure what we have established so far:

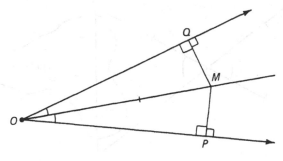

Figure 7.13

We have now established the **ASA** criterion, and we conclude that

$$\triangle OQM \simeq \triangle OPM.$$

Example. In isosceles triangle $\triangle ABC$, $|AB| = |BC|$, and $\overline{BN}$ bisects $\angle ABC$. Prove that $\triangle ABN \simeq \triangle CBN$.

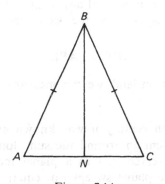

Figure 7.14

Proof. We are given that $|AB| = |BC|$. By the definition of angle bisector, we have that $m(\angle ABN) = m(\angle CBN)$. Both triangles share side $\overline{BN}$. We therefore satisfy the conditions for the **SAS** congruence axiom, and we conclude that $\triangle ABN \simeq \triangle CBN$.

Example. A surveyor wishes to measure the width of a lake without getting his feet wet. He proceeds as follows. Pick two points A, B on opposite sides of the lake. We want to measure the distance between A and B. Pick a point C on one side, such that the path from C to A and C to B does not go over the water, as shown on the figure.

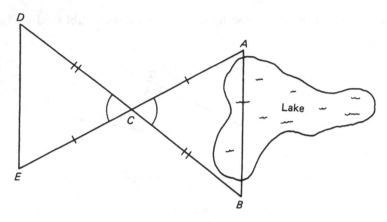

Figure 7.15

We can then measure $\overline{CA}$ and $\overline{CB}$. We now go along the line L_{BC} to the point D so that $|CD| = |CB|$. We go along the line L_{AC} to the point E so that $|CE| = |CA|$. Then the triangles

$$\triangle DCE \quad \text{and} \quad \triangle BCA$$

are congruent by **SAS**: Their vertical angles at C have the same measure, and the two adjacent sides have the same measure by construction. Therefore

$$|DE| = |AB|.$$

Since the segment $\overline{DE}$ is on land, we can measure it, thus giving also the length $|AB|$.

Example. By the 17th century it was known experimentally that the planets in our solar system go around the sun along curved orbits called ellipses. It was also observed that as a planet moves, the line segment between the sun and the planet sweeps out equal areas in equal time as illustrated in Figure 7.16(a).

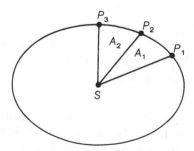

Figure 7.16(a)

In the figure, S is the sun, and P_1 is the position of the planet at a certain time. Let P_2 be the planet's position after one day; let P_3 be its position after another day. The areas of regions A_1 and A_2 are equal. A major question confronting scientists at that time was the nature of the physical laws that governed this motion.

In the 1660's, while still in his twenties, Englishman Isaac Newton proposed the theory of gravity: a single universal force which keeps the planets in motion around the sun, the moon in orbit around the earth, and which causes apples to fall from the tree to the ground. Among other things, Newton showed a connection between the fact that the force acting on a planet is towards the sun, and the fact that the areas are equal. In this example, we shall reproduce some of Newton's reasoning, which uses the ideas about areas and congruences of triangles discussed in this chapter to approximate these areas.

Suppose the sun did not exert any force on the planet. Then the planet would be moving on a straight line at uniform speed. Let the planet be at position P_1 at a certain time. After one second, let the planet be at P_2, and after another second let the planet be at P_3, as shown on Figure 7.16(b).

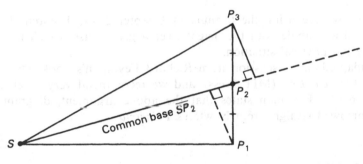

Figure 7.16(b)

Let S be the sun. Then $|P_1P_2| = |P_2P_3|$. The triangles $\triangle SP_1P_2$ and $\triangle SP_2P_3$ have a common base $\overline{SP_2}$. By using **ASA**, show that their heights with respect to this base are equal. It then follows that their areas are equal, that is

$$\text{area of } \triangle SP_1P_2 = \text{area of } \triangle SP_2P_3.$$

However, the sun exerts a force on the planet, and this force is directed toward the sun. We shall draw an average approximation. The effect of the attraction of the sun is to change the motion by some amount in the direction of a line parallel to $\overline{SP_2}$. This means that the planet starting from P_2 after one second is located at P_4, lying on the line through P_3 parallel to $\overline{SP_2}$ as shown on Figure 7.16(c).

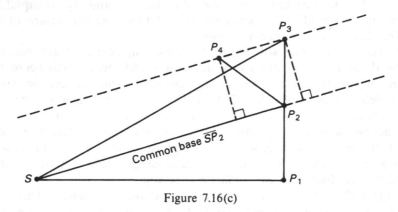

Figure 7.16(c)

But then again triangle $\triangle SP_2P_4$ has the same base $\overline{SP_2}$, and the same height with respect to this base, so

$$\text{area of } \triangle SP_2P_4 = \text{area of } \triangle SP_2P_3 = \text{area of } \triangle SP_1P_2.$$

Note how we are using the formula of Chapter 3, §1, Theorem 3-2 for the area of a triangle with the height over a given base, which is forced on us by the physical situation.

We adapted this discussion from Richard Feynman's book *The Character of Physical Law* (MIT Press), and we recommend very much reading this book. Feynman states that the above argument, diagram and all, is borrowed straight from Newton's *Principia*.

7, §1. EXERCISES

1. Given that $\triangle ABC \simeq \triangle PQR$. Write down everything you can conclude from this fact.

2. Let ABC and XYZ be triangles. In each situation below, state whether or not $\triangle ABC \simeq \triangle XYZ$, and give your reasons. Draw pictures!
   (a) $|AB| = |ZX|$, $|BC| = |XY|$, $|AC| = |ZY|$
   (b) $|AB| = |XY|$, $|AC| = |XZ|$, $m(\angle B) = m(\angle Y)$
   (c) $m(\angle A) = m(\angle X)$, $m(\angle B) = m(\angle Y)$, $m(\angle C) = m(\angle Z)$
   (d) $m(\angle A) = m(\angle X)$, $m(\angle B) = m(\angle Y)$, $|AC| = |XZ|$
   (e) $m(\angle B) = m(\angle Y)$, $m(\angle C) = m(\angle Z)$, $|BC| = |XZ|$
   (f) $|BC| = |ZY|$, $|CA| = |YX|$, $m(\angle C) = m(\angle Y)$

3. In isosceles triangle $\triangle ABC$ below, $|AB| = |BC|$. Segment $\overline{BN}$ is drawn perpendicular to $\overline{AC}$.

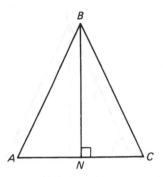

Figure 7.17

Prove that $\overline{BN}$ bisects $\angle ABC$.

4. Prove the following corollary to **ASA**. Given right triangles $\triangle ABC$ and $\triangle XYZ$, with right angles B and Y. Assume

$$|BC| = |YZ|$$

and

$$m(\angle A) = m(\angle X).$$

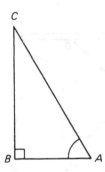 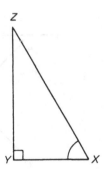

Figure 7.18

Prove that $\triangle ABC \simeq \triangle XYZ$.

5. Use the result of Exercise 4 to prove the **converse of the Isosceles Triangle Theorem** (Theorem 5-3), namely:

Given $\triangle PQR$, with $m(\angle P) = m(\angle R)$.

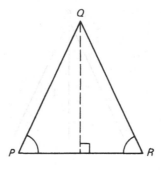

Figure 7.19

Prove: $|PQ| = |QR|$. [*Hint*: Draw a line segment from Q perpendicular to $\overline{PR}$.]

6. Use the results of Exercise 5 to prove that a triangle with three equal-sized angles (i.e. 60° each) is equilateral. This is the converse of the Corollary 5-3.

7. The figure shows a side view of an ironing board. So that the board may be positioned at several different heights, the legs are hinged at H, and point B is free to slide along a track on the underside. In addition, $|CH| = |HB|$ and $|AH| = |HD|$. Using congruent triangles, explain how this design ensures the fact that, no matter what height the board is set to, it will always remain horizontal.

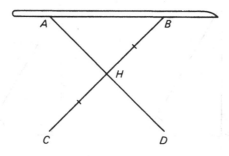

Figure 7.20

8. In order to measure the distance from the shore (point A) to a ship at sea (point D), the Greek mathematician Thales devised the instrument shown in the picture. He held rod AB vertically and sighted point D along rod BC. Then, keeping $\angle CBA$ constant, he revolved the instrument around AB so that it pointed in to shore. Sighting along BC', he then located point D'. Using congruent triangles, explain how the device worked.

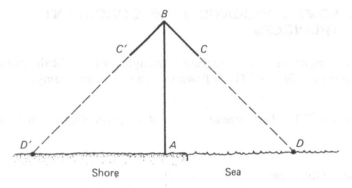

Figure 7.21

9. Another method used by Thales to measure the distance from shore (point X) to ship (point Y) is pictured below. Picking another point Z along the shore, he measured ∠ZXY and ∠XZY and reproduced them in ∠ZXY′ and ∠XZY′ respectively. Using congruent triangles, explain how this method worked.

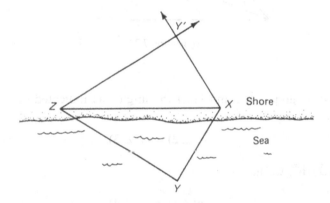

Figure 7.22

10. The pliers in the picture are hinged at points B, C and E; and slotted at points A and D. In addition, |AE| = |EC| and |BE| = |ED|. Using congruent triangles, explain why the jaws will always remain parallel.

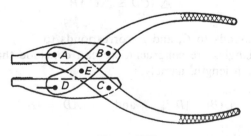

Figure 7.23

7, §2. SOME APPLICATIONS OF CONGRUENT TRIANGLES

We can sometimes use congruent triangles to establish properties of more complex figures. The following theorem is an example.

Theorem 7-1. *The opposite sides of a parallelogram have the same length.*

Proof. Given parallelogram *ABCD*:

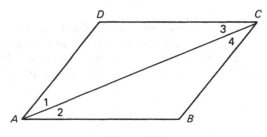

Figure 7.24

We draw diagonal $\overline{AC}$, and label the angles as indicated. Since $\overline{AB}\|\overline{DC}$, we have

$$m(\angle 2) = m(\angle 3).$$

Since $\overline{AD}\|\overline{BC}$, we have

$$m(\angle 1) = m(\angle 4).$$

Both triangle $\triangle ACD$ and triangle $\triangle ACB$ share $\overline{AC}$ as a common side. Therefore, by the **ASA** congruence property, we have

$$\triangle ACD \simeq \triangle CAB$$

where *A* corresponds to *C*, and *D* corresponds to *B*.

Since the triangles are congruent, we can conclude that corresponding sides are equal in length, namely:

$$|AB| = |DC| \quad \text{and} \quad |AD| = |BC|.$$

This proves the theorem.

Theorem 7-2. *A parallelogram of height h and base b has*

$$area = bh.$$

Proof. We use again the decomposition as in Theorem 7-1 for the parallelogram into two triangles, as illustrated on the figure.

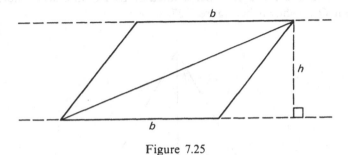

Figure 7.25

Each triangle has base b and height h. Hence each triangle has area $\frac{1}{2}bh$. Hence the area of the parallelogram is equal to the sum of the areas of the two triangles, which is

$$\tfrac{1}{2}bh + \tfrac{1}{2}bh = bh,$$

thus proving the theorem.

The above proof followed the ideas of Theorem 7-1, and also followed the pattern given for the proof of the area of a trapezoid. There is however an additional interesting symmetry. The formula is the same as that for the area of a rectangle with sides of lengths b and h.

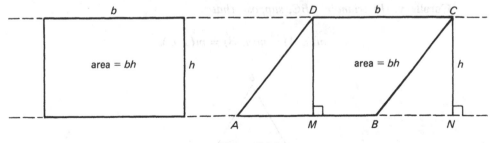

Figure 7.26

Exercise. In the figure, let $\overline{DM}$ and $\overline{CN}$ be the perpendicular line segments as shown. Give another proof for Theorem 7-2 by proving that triangles $\triangle ADM$ and $\triangle BCN$ are congruent. Observe that

area of parallelogram $ABCD$ = area of $MBCD$ + area of triangle.

We can now also prove the converse of the Isosceles Triangle Theorem given in §2 of Chapter 5:

Theorem 7-3. *In triangle* $\triangle ABC$, *assume that* $m(\angle B) = m(\angle C)$. *Then the sides oppsite B and C have the same length.*

Proof. The line through A perpendicular to $\overline{BC}$ intersects the line $\overline{BC}$ in a point Q as shown in Figure 7.27.

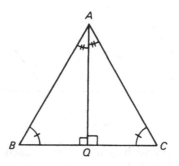

Figure 7.27

The two triangles $\triangle AQB$ and $\triangle AQC$ have a side in common, namely $\overline{AQ}$. The angles $\angle AQB$ and $\angle AQC$ are right angles. The angles $\angle B$ and $\angle C$ have the same measure. Hence the angles $\angle BAQ$ and $\angle QAC$ have the same measure (why?). Therefore the triangles $\triangle AQB$ and $\triangle AQC$ satisfy **ASA**, and are therefore congruent. If follows that $|AB| = |AC|$, thus proving the theorem.

Corollary. *In triangle ABC, suppose that*

$$m(\angle A) = m(\angle B) = m(\angle C).$$

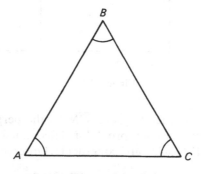

Figure 7.28

Then triangle ABC is equilateral, in other words

$$|AB| = |BC| = |CA|.$$

Proof. We apply the preceding theorem two times. Since $m(\angle A) = m(\angle B)$, we have $|AC| = |BC|$. Since $m(\angle B) = m(\angle C)$, we have $|AB| = |AC|$. We conclude that all three sides have the same length.

The proofs of the next two theorems will be left as exercises (see Exercises 3 and 4).

Theorem 7-4. *If the opposite sides of a quadrilateral have the same length (i.e. each opposite pair have the same length), then the quadrilateral is a parallelogram.*

Theorem 7-5. *If one pair of opposite sides of a quadrilateral are equal in length and parallel, then the quadrilateral is a parallelogram.*

7, §2. EXERCISES

Parallelograms

1. Given parallelogram $ABCD$, with diagonals intersecting at O:

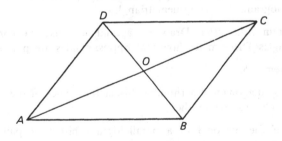

Figure 7.29

Show that

$$|AO| = |OC|$$

and

$$|BO| = |OD|$$

using Theorem 7-1 and the **ASA** congruence property. This result is often stated as a theorem:

The diagonals of a parallelogram bisect each other.

2. Given quadrilateral MATH. If the diagonals bisect each other, prove that MATH must be a parallelogram.

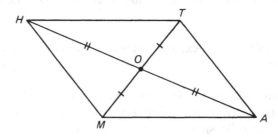

Figure 7.30

Given:

$$|MO| = |OT|,$$

$$|HO| = |OA|.$$

Of course, this is the converse of Exercise 1.

Exercise 2 provides a test of whether a quadrilateral is a parallelogram. The next two exercises give two more tests. In each, you must use what is given to prove that the opposite sides of the quadrilateral are parallel, which is the definition of a parallelogram. Use congruent triangles.

3. Prove Theorem 7-4. [*Hint*: Draw a diagonal, and use the converse of the Alternate Angles Theorem to show that opposite sides are parallel.]

4. Prove Theorem 7-5.

5. Prove that the diagonals of a rhombus bisect the angles of the rhombus, and are perpendicular to one another.

6. Prove that if the diagonal of a parallelogram bisects a pair of opposite angles, then the parallelogram is a rhombus.

Angle Bisectors and Inscribed Circle

7. Point O is on the bisector of $\angle PQR$. Prove that O is equidistant from the sides of the angle. In other words, if the perpendicular from O to $\overline{PQ}$ intersects $\overline{PQ}$ at X, and the perpendicular from O to $\overline{QR}$ intersects $\overline{QR}$ at Y, then

$$|OX| = |OY|.$$

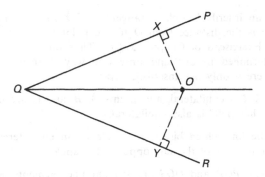

Figure 7.31

8. Prove the converse of the statement in Exercise 7. In other words prove:

Let O be a point inside an acute angle, and assume that O is equidistant from the sides of the angle. Prove that O lies on the bisector of the angle.

9. (a) Prove the following theorem.

Theorem on the Angle Bisectors of a Triangle. *Let $\triangle PQA$ be a triangle. Let L_1, L_2, L_3 be the three lines which bisect the three angles of the triangle, respectively. Let O be the point of intersection of L_1 and L_2. Then O lies on L_3.*

[*Hint*: Use the results of Exercises 7 and 8.]

Let T be a triangle. By an **inscribed** circle, we mean a circle such that the three sides of the triangle are tangent to the circle.

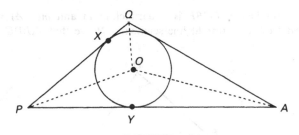

Figure 7.32

An inscribed circle always exists. Namely, let r be the distance from O to $\overline{PQ}$, so $r = d(O, X)$, where X is the point of intersection of $\overline{PQ}$ with the line through O perpendicular to $\overline{PQ}$. Then by Exercise 8 and the previous theorem r is also the distance from O to $\overline{QA}$ and to $\overline{PA}$, so the circle of center O and radius r is an inscribed circle. Is there another inscribed circle? The answer is no.

(b) Let C be an inscribed circle to the triangle $\triangle PQA$, and let M be the center of C. Prove that M is the point of intersection of the angle bisectors of the triangle.

Since an inscribed circle is tangent to $\overline{PQ}$, for instance, the radius of the circle is the distance from O to the point where the circle intersects $\overline{PQ}$ (see Exercise 8 of Chapter 5, §3). Thus for any inscribed circle we have determined in one and only one way its center and its radius. Hence there is only one inscribed circle.

10. Triangle $\triangle XYZ$ is equilateral, and points A, B, and C are midpoints of the sides. Prove that ABC is also equilateral.

11. Prove that the line which bisects an angle of an equilateral triangle is the perpendicular bisector of the side opposite the angle.

12. In the diagram, POR and QOS are straight line segments, $m(\angle 1) = m(\angle 2)$ and $m(\angle 3) = m(\angle 4)$. Prove that O is the midpoint of $\overline{QS}$. [*Hint*: First get $\triangle PQR \simeq \triangle PSR$. Then find another pair of congruent triangles to get $|SO| = |OQ|$.]

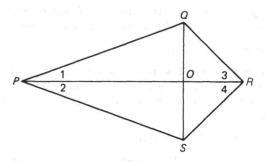

Figure 7.33

13. In the figure below, $ECBF$ is a parallelogram and $m(\angle A) = m(\angle D)$. Also EFA and DCB are straight line segments. Prove that $\triangle EDC \simeq \triangle FAB$.

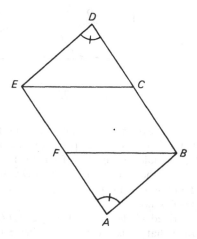

Figure 7.34

14. The procedure for copying an angle (which we used in Construction 1-3) is as follows.

Given an angle $\angle AOB$. Draw a ray, labeling its endpoint P:

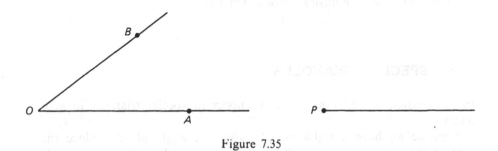

Figure 7.35

With the compass tip on O draw an arc intercepting the two rays of $\angle AOB$ in points S and T. Using the same setting, draw an arc with the compass tip on P; intersecting the ray in point Q.

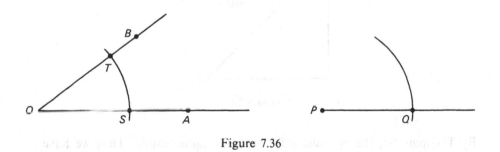

Figure 7.36

Set the compass tips at distance $d(S, T)$. With the point of the compass on Q, draw an arc intercepting the previously drawn arc in point R.

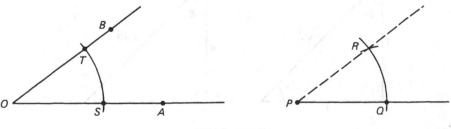

Figure 7.37

The ray from point P through R gives rise to the angle $\angle RPQ$. Use congruent triangles to show that

$$m(\angle O) = m(\angle P).$$

This is one way of justifying Construction 1-3.

7, §3. SPECIAL TRIANGLES

In this section we discuss the most important special triangles in geometry.

Suppose we have a right triangle with one angle of 45°. Since the sum of the angles measures 180°, it follows that the other acute angle also has 45°.

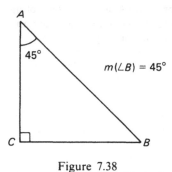

Figure 7.38

By Theorem 7-3, the opposite sides have the same length. Thus we have an isosceles right triangle, which we studied in §2 of Chapter 5. Recall that if an isosceles right triangle has legs of length a, then the hypotenuse has length $a\sqrt{2}$:

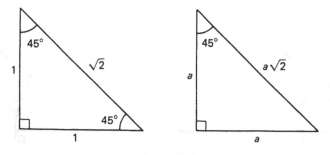

Figure 7.39

We can summarize all of this in a theorem:

Theorem 7-6. *If a right triangle has an angle of 45°, it is an isosceles right triangle. Let a be the length of the legs. Then $a\sqrt{2}$ is the length of the hypotenuse. Hence the sides stand in the ratio $1:1:\sqrt{2}$.*

Example. Find the perimeter of the triangle below.

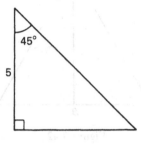

Figure 7.40

Since the triangle has one angle of 45°, we can conclude that both legs have length 5, and that the hypotenuse has length $5\sqrt{2}$. The perimeter therefore is

$$5 + 5 + 5\sqrt{2} = 10 + 5\sqrt{2}.$$

Theorem 7-7. *In a right triangle with acute angles of 30° and 60°, the side opposite the 30° angle is one half the length of the hypotenuse. The sides stand in the ratio*

$$1:2:\sqrt{3}.$$

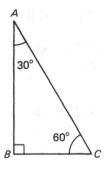

Figure 7.41

Proof. Given right triangle $\triangle ABC$ as illustrated above. We wish to show that

$$|BC| = \tfrac{1}{2}\cdot|AC|.$$

Let C' be the point on line L_{BC} such that $|C'B| = |BC|$ and $C' \neq C$.

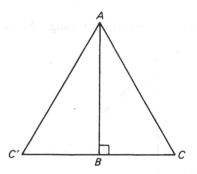

Figure 7.42

Then triangles $\triangle AC'B$ and $\triangle ACB$ are congruent by **SAS**: They have a common side $\overline{AB}$, we have $|C'B| = |BC|$ by construction, and the angles $\angle C'BA$ and $\angle CBA$ are right angles. By **SAS** we conclude:

$$\triangle ABC \simeq \triangle ABC'$$

and therefore

$$m(\angle C'AB) = m(\angle CAB) = 30°,$$
$$m(\angle C') = m(\angle C) \quad = 60°,$$

and

$$|CB| = |C'B|.$$

This means that $m(\angle C) = m(\angle C') = m(\angle CAC') = 60°$. By the corollary proved earlier, we can conclude that triangle $\triangle ACC'$ is equilateral. In particular,

$$|CC'| = |AC|.$$

Since $|CB| = |C'B|$, we have

$$|CB| = \tfrac{1}{2}|AC|.$$

If $a = |CB|$ then $|AC| = 2a$. By Pythagoras, we find the length of the third side,

$$|AB|^2 = (2a)^2 - a^2 = 4a^2 - a^2 = 3a^2,$$

whence

$$|AB| = \sqrt{3}\,a.$$

This shows that the sides have lengths a, $2a$, $\sqrt{3}\,a$ respectively, and concludes the proof of the theorem.

A triangle whose angles have measures 30°, 60°, 90° is called a **30-60-90 right triangle**, or just a **30-60 right triangle**.

Example. Suppose the short leg of a 30-60-90 right triangle has length 1. Then the hypotenuse has length 2, and by Pythagoras, the other leg has length $\sqrt{4-1} = \sqrt{3}$. This is illustrated by Figure 7.43(a). If the short leg has length a, then the hypotenuse has length $2a$, and the other leg has length $a\sqrt{3}$, as on Figure 7.43(b).

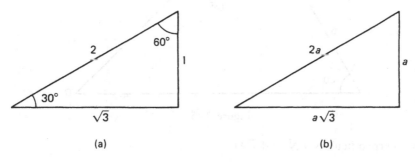

(a)　　　　　　　　　　　　(b)

Figure 7.43

Example. In a 30-60 right triangle, suppose the short leg has length 5 cm. Then the hypotenuse has length 10 cm, and the other leg has length $5\sqrt{3}$ cm.

Example. A man is flying a kite. He has let out 50 m of string, and he notices that the string makes an angle of 60° with the ground. How high is the kite?

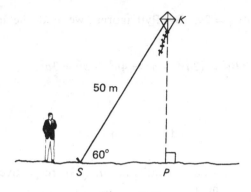

Figure 7.44

Since the angle at S has $60°$, the angle at K must have $30°$, and we have a 30–60–90 triangle. Thus

$$|SP| = \tfrac{1}{2} \cdot |SK| = 25 \text{ m}.$$

Hence the height of the kite is $25\sqrt{3}$ m.

Example. Find the area of the illustrated trapezoid:

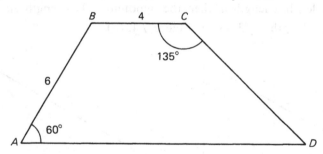

Figure 7.45

Draw perpendiculars $\overline{BN}$ and $\overline{CM}$.

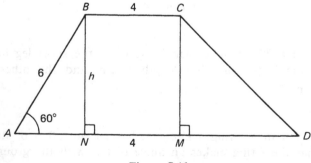

Figure 7.46

Triangle $\triangle ABN$ is a 30–60–90 triangle, and therefore

$$|AN| = \tfrac{1}{2} \cdot 6 = 3.$$

If we let $|BN| = h$, we have

$$h^2 + 3^2 = 6^2 \qquad \text{applying Pythagoras to } \triangle ABN,$$
$$h^2 + 9 = 36,$$
$$h^2 = 27,$$
$$h = \sqrt{27}.$$

By our **PD** property, we have that

$$|CM| = |BN| = \sqrt{27}.$$

Since $\angle BCM$ is a right angle,

$$m(\angle DCM) = 135° - 90°$$
$$= 45°.$$

Therefore, $\triangle DCM$ is an isosceles right triangle, and we have

$$|CM| = |DM| - \sqrt{27}.$$

We now have the lengths of the two bases and the height of the trapezoid:

$$\text{upper base:} \quad 4,$$
$$\text{lower base:} \quad 3 + 4 + \sqrt{27},$$
$$\text{height:} \quad \sqrt{27}.$$

We compute the area using the formula

$$\text{area} = \tfrac{1}{2}(b_1 + b_2)h$$
$$= \tfrac{1}{2}(4 + 7 + \sqrt{27})\sqrt{27}$$
$$= \tfrac{1}{2}(11\sqrt{27} + 27) \text{ sq. units.}$$

Advice. Right triangles of the above types are so important in mathematics that you should memorize the properties we have just discussed, just as you memorized the multiplication table. This memorization should be done by repeating out loud the following figures, like a poem:

> *Forty five, forty five; one, one, square root of two.*
> *Thirty, sixty; one, two, square root of three.*

7, §3. EXERCISES

1. Prove that $\triangle ABC \simeq \triangle XYZ$.

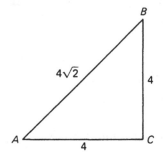

Figure 7.47

2. In the figure, the circle centered at O has radius 8. If $m(\angle ROT) = 120°$, find the area of triangle $\triangle ROT$. [*Hint:* Draw the height from O to $\overline{TR}$. Since $m(\angle T) = M(\angle R) = 30°$, you've created two 30-60-90 triangles.]

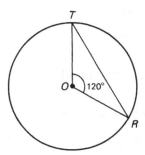

Figure 7.48

3. Find the area of the trapezoid illustrated below:

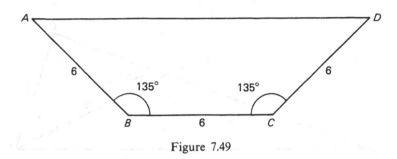

Figure 7.49

[*Hint*: Draw perpendiculars from B and C to $\overline{AD}$. Consider the area of the rectangle, plus the area of the two right triangles.]

4. What is the area of the following trapezoid?

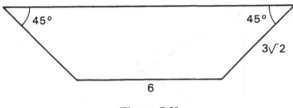

Figure 7.50

5. The roof in your attic is pitched at $60°$, as shown. You wish to add a room up there, with vertical side walls 3 m high. If the attic floor is 6 m wide, how far away from the sides must you build the walls?

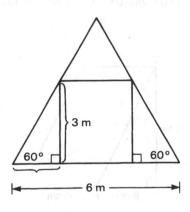

Figure 7.51

6. Find the area of triangle $\triangle JOY$.

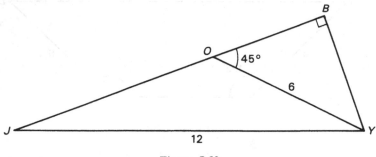

Figure 7.52

7. Given $\triangle ABC$ is equilateral. Prove $|BC| = |CD|$.

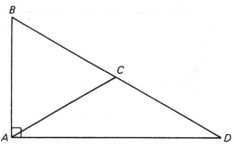

Figure 7.53

8. In the figure below $\overline{UI} \parallel \overline{QE}$ and $\overline{QU} \parallel \overline{EI}$. Find the area of quadrilateral $QUIE$ and the length of $\overline{UI}$.

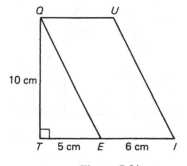

Figure 7.54

9. In the figure below $\overline{XM} \parallel \overline{SA}$, $\overline{MA} \parallel \overline{XS}$, and $\overline{MO} \perp \overline{OS}$. Find the area of quadrilateral $XMAS$.

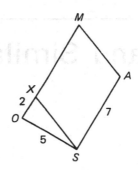

Figure 7.55

CHAPTER 8

Dilations and Similarities

8, §1. DEFINITION

Consider the pair of quadrilaterals as on Figure 8.1:

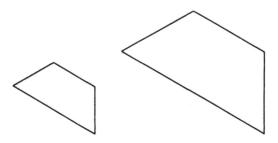

Figure 8.1

Obviously, the two quadrilaterals are not congruent, but they appear related in a certain way. We would like to say that they have a "similar" shape. How can we describe the way we can obtain one of these quadrilaterals from the other?

Let us start with a special case of the general concept, which will be called a **dilation**.

Let O be a given point in the plane. To each point P of the plane we associate the point P' which lies on the ray R_{OP} at a distance from O equal to *twice* that of P from O. The point P' in this case is also denoted by $2P$. See Figure 8.2.

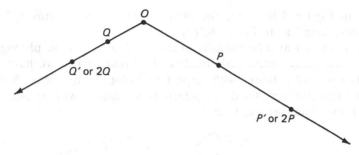

Figure 8.2

The association which to each point P associates the point $2P$ (or P') is called **dilation by 2**, relative to O. We shall write the point $2P$ also with the notation

$$D_{2,O}(P) = 2P.$$

If we have agreed to keep the origin O fixed, we do not need to mention it in the notation, and we write simply

$$D_2(P) = 2P.$$

We can dilate by other numbers besides 2. In fact, let r be any positive number ($r > 0$). We define the **dilation by r** relative to a point O in a manner similar to dilation by 2. To each point P in the plane, we associate the point $D_{r,O}(P)$, which is the point on ray R_{OP} which is r times the distance from O as P is. Again, if we do not mention the point O in the notation, we simply write

$$D_r(P) = rP.$$

In Figure 8.3(a) we have drawn the points Q, $D_2(Q)$, and $D_3(Q)$.
In Figure 8.3(b) we have drawn the points P, $D_{1/2}(P)$, and $D_{1/4}(P)$.

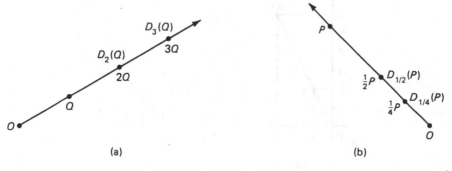

Figure 8.3

Notice in Figure 8.3(b) that the plane seems to be "shrinking" rather than "stretching" as in Figure 8.3(a).

A dilation can also be thought of as a "blow-up" as in photography, or as a "shrinking" as in scale models. In Figure 8.4(a) we have shown the "blow-up" of a flower with respect to point O. In Figure 8.4(b) we show a "scaled-down" transistor which is a dilation with respect to the point O, by a number less than 1.

(a)   (b)

Figure 8.4

We shall now see how to describe a dilation in terms ⌄i coordinates.

Let $A = (a_1, a_2)$ be a point of the plane, given in terms of coordinates. Let r be a positive number. We define the **product** rA to be the point

$$rA = (ra_1, ra_2).$$

We have multiplied each coordinate of A by r to get the coordinate of rA.

Example. Let $A = (1, 2)$ and $r = 3$. Then $3A = (3, 6)$, as shown on Figure 8.5(a).

Example. Let $A = (1, 2)$ and $r = \frac{1}{2}$. Then

$$\tfrac{1}{2}A = (\tfrac{1}{2}, 1)$$

as shown on Figure 8.5(b).

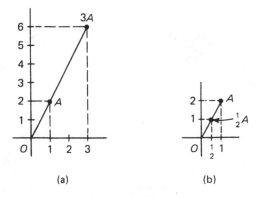

(a)   (b)

Figure 8.5

Geometrically, we see that multiplication by 3 dilates A by 3 with respect to the origin. On the other hand, dilating by $\frac{1}{2}$ amounts to halving, e.g. in Figure 8.5(b).

Example. Let $A = (-2, 3)$ and $r = 2$. Then $rA = (-4, 6)$.

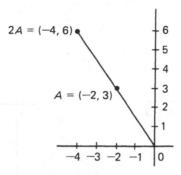

Figure 8.6

In each example, we see from the picture that rA is the dilation of A by a factor of r. We can also prove algebraically that the distance of rA to the origin is equal to r times the distance of A to the origin. In fact, we have

$$d(O, rA) = \sqrt{(ra_1)^2 + (ra_2)^2}$$
$$= \sqrt{r^2 a_1^2 + r^2 a_2^2}$$
$$= \sqrt{r^2(a_1^2 + a_2^2)}$$
$$= \sqrt{r^2}\sqrt{a_1^2 + a_2^2}$$
$$= r \cdot d(O, A).$$

It is natural now to define the product cA for any real number c, not just a positive number, to be the point

$$cA = (ca_1, ca_2).$$

Thus we multiply each coordinate of A by c to get the coordinates of cA. We shall discuss examples, and the geometric interpretation, which will be used further in Chapters 10 and 11. You can skip the rest of this section now if you are interested only in dilations and their applications.

Example. Let $A = (2, 5)$ and $c = -6$. Then $cA = (-12, -30)$.

Example. Let $A = (-3, 7)$ and $c = -4$. Then $cA = (12, -28)$.

Note that in this case we have chosen c to be negative.

Example. Take $A = (1, 2)$ and $c = -3$. We write $-A$ instead of $(-1)A$. We have drawn $-A$ and $-3A$ in Figure 8.7(a) and (b).

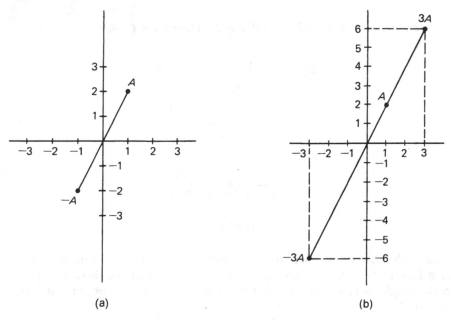

(a)   (b)

Figure 8.7

If c is negative, we write $c = -r$, where r is positive, and we see that multiplication of A by c can be obtained by first multiplying A by r and then taking the reflection $-rA$. Thus we can say that $-rA$ points in the opposite direction from A, with a stretch of r. We shall study reflections more systematically in Chapter 11, §5.

8, §1. EXERCISES

1. On your paper, draw a point O, and some other points P, Q, R, as shown:

Locate and label each of the following points, representing dilations with respect to point O:

(a) $2P$ (b) $3Q$ (c) $\frac{1}{3}R$ (d) $1P$ (e) $\frac{1}{2}P$

2. Is there any value of r such that $D_r(P) = P$ for all points P?

3. Using a ruler, draw a triangle ABC on your paper. Choose a point O inside the triangle. Dilate each point A, B, and C by 2 with respect to O. Connect the points $2A$, $2B$, $2C$.

4. Using the same triangle ABC as in the previous exercise, locate a point O *outside* the triangle. Dilate each point A, B, and C by 2 with respect to this new point O. Connect the points $2A$, $2B$, $2C$. How does this dilated triangle compare with the one created in the previous exercise?

5. Write the coordinates for cA with the following values of c and A. In each case, draw A and cA.
   (a) $A = (-3, 5)$ and $c = 4$ (b) $A = (4, -2)$ and $c = 3$
   (c) $A = (-4, -5)$ and $c = 2$ (d) $A = (2, -3)$ and $c = 2$
   (e) $A = (3, 2)$ and $c = \frac{1}{2}$ (f) $A = (3, 2)$ and $c = \frac{1}{3}$
   (g) $A = (4, 3)$ and $c = \frac{1}{2}$ (h) $A = (5, 3)$ and $c = \frac{2}{3}$

6. Again write the coordinates for cA with the following values of c and A. In each case, draw A and cA.
   (a) $A = (-3, 5)$ and $c = -4$ (b) $A = (4, -2)$ and $c = -3$
   (c) $A = (-4, -5)$ and $c = -2$ (d) $A = (2, -3)$ and $c = -2$
   (e) $A = (3, 2)$ and $c = -\frac{1}{2}$ (f) $A = (3, 2)$ and $c = -\frac{1}{3}$
   (g) $A = (4, 3)$ and $c = -\frac{1}{2}$ (h) $A = (5, 3)$ and $c = -\frac{2}{3}$
   Compare the results of Exercises 5 and 6 on the same sheet of paper.

7. In each case copy the rectangle onto a piece of graph paper. Then draw the rectangle obtained by dilating by 2 with respect to the origin. Write down the area of the original rectangle and the area of the dilated rectangle.

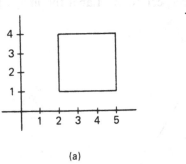   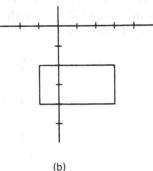

(a)   (b)

Figure 8.8

8. Copy the rectangles below onto graph paper, then draw the rectangles obtained by dilating by 1.5 with respect to the origin. Write down the area of the original rectangle and the area of the dilated rectangle.

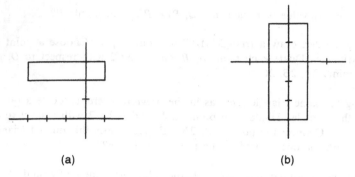

(a)　　　　　　　　(b)

Figure 8.9

EXPERIMENT 8-1

1. Look at the right triangle illustrated below:

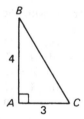

Figure 8.10

(a) How long is the hypotenuse of this triangle?
(b) Draw the triangle on your paper (a sketch will do). Dilate the points C and B by 2 with respect to A. Label the image points C' and B', as shown:

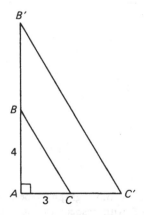

Figure 8.11

(c) How long are the legs $\overline{AC'}$ and $\overline{AB'}$ of the new triangle?

(d) How long is the hypotenuse of the new triangle? What is the ratio of this length to the original length?

(e) What is the area of the new triangle? What is the area of the original triangle? What is the ratio of the areas?

(f) How does the measure of $\angle C$ compare with the measure of $\angle C'$?

2. Construct a right triangle with legs 5 cm and 8 cm. Use a ruler, straightedge, protractor, etc. to get an accurate picture. Choose a point O not on the triangle, as shown:

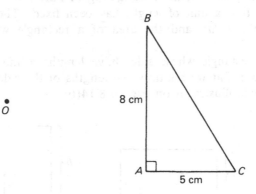

Figure 8.12

(a) Dilate each point A, B, and C by 2 with respect to O. Use a ruler to get an accurate picture. Label the image points A', B', and C'.

(b) Is triangle $A'B'C'$ another right triangle?

(c) By measuring, find the length of the hypotenuse of $\triangle A'B'C'$. Measure the lengths of the legs and compute the area of $\triangle A'B'C'$.

(d) Measure the angles of $\triangle ABC$ and measure the angles of $\triangle A'B'C'$. How do the measurements compare?

3. Accurately draw a rectangle with sides of length 2 cm and 3 cm as shown:

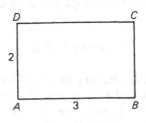

Figure 8.13

(a) Dilate the points A, B, C, and D by 2.5 with respect to a point O not on the rectangle.

(b) Compare the areas and perimeters of the two rectangles.

8, §2. CHANGE OF AREA UNDER DILATION

What happens to the area of a figure when the figure is dilated? We begin answering this question by considering a simple case: the rectangle. You have already done some investigating in Part 3 of Experiment 8-1.

We assume that a unit of length has been fixed. Then the area of a square of side a is a^2, and the area of a rectangle whose sides have lengths a, b is ab.

Consider a rectangle whose sides have lengths a and b as on Figure 8.14(a). Suppose that we multiply the lengths of the sides by 2, and obtain the rectangle illustrated on Figure 8.14(b).

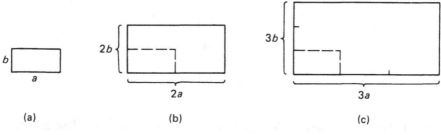

(a)   (b)   (c)

Figure 8.14

Then the sides of this dilated rectangle have lengths $2a$ and $2b$. Hence the area of the dilated rectangle is equal to $2a2b = 4ab = 2^2ab$. Similarly, suppose that we dilate the sides by 3, as illustrated on Figure 8.14(c). Then the sides of the dilated rectangle have lengths $3a$ and $3b$, whence the area of the dilated rectangle is equal to $3a3b = 9ab = 3^2ab$.

In general, let S be a rectangle whose sides have lengths a, b respectively. Let rS be the dilation of S by r. Then the sides of rS have lengths ra and rb respectively, so that the area of the dilated rectangle is equal to

$$(ra)(rb) = r^2ab.$$

Thus the area of rectangles changes by r^2 under dilation by r.

Consider what happens when r is less than 1, say $r = \frac{1}{2}$, for example. If we dilate a rectangle by $\frac{1}{2}$, then the area of the dilated rectangle is $\frac{1}{4}$ as large as the original ($\frac{1}{4} = (\frac{1}{2})^2$). Figure 8.15 illustrates this fact.

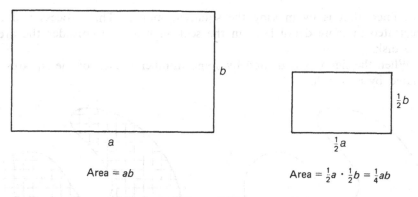

Area = ab   Area = $\frac{1}{2}a \cdot \frac{1}{2}b = \frac{1}{4}ab$

Figure 8.15

We can use this information about rectangles to see what happens to the areas of arbitrary figures in the plane when they are dilated. In Figure 8.16 we illustrate an arbitrary region S.

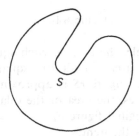

Figure 8.16

We can approximate the area of this figure by drawing a fine grid of squares, very much like those you see on graph paper:

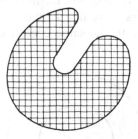

Figure 8.17

We then count up the number of squares inside the figure, and multiply this number by the area of one of the squares. This gives us an approximate area for S. We can make the approximation better by making the grid finer, that is by making the squares smaller. This process will be illustrated in more detail later in the section when we consider the area of a disk.

When the figure S is dilated by some number r, each of the squares is dilated by r as well.

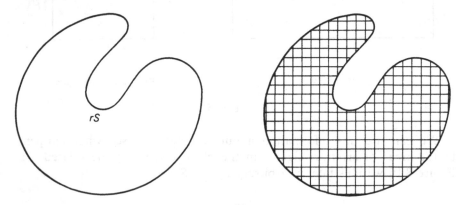

Figure 8.18

Since a square is a rectangle, the area of each square is multiplied by r^2. Therefore the sum of the areas of all the squares is multiplied by r^2. Since the area of the dilated figure rS is approximated by the sum of the areas of these dilated squares, the area of the dilated figure is r^2 times as large as the area of the original figure S.

We state this result as a theorem:

Theorem 8-1. *Let S be an arbitrary region in the plane with area A. Let rS be the image of S under a dilation by a positive number r. Then the area of rS is $r^2 \cdot A$.*

We apply this theorem to find the formula for the area of a disk of radius r. Let D_1 be the disk of radius 1, and let D_r be the disk of radius r. If we draw D_1 and D_r with the same center, we can easily see that D_r is the dilation of D_1 by r:

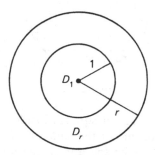

Figure 8.19

By Theorem 8-1 the area of D_r is r^2 times the area of D_1. We let π denote the numerical value of the area of D_1. Therefore:

$$\text{the area of } D_r \text{ is } r^2\pi.$$

Of course there remains the problem of determining the value of π, the area of D_1. There are various devices to do this, and it can be shown that π is equal to a decimal, which to five places is given by

$$\pi = 3.14159\ldots .$$

In §4 we shall prove that π satisfies another relation, namely π is the ratio of the circumference to the diameter of a circle:

$$c = \pi d \quad \text{and} \quad \pi = c/d,$$

where c is the circumference and d is the diameter. Thus the π we are using now is the same one that you may already know. This relationship between the circumference and diameter of a circle gives us a method for computing π, as follows.

EXPERIMENT 8-2

Take a circular pan. Measure the circumference and diameter with a soft measuring tape. Take the ratio. This will give you a value for π, good at least to one decimal place.

There are more sophisticated ways of finding more decimal places for π. Those which give π with arbitrarily good accuracy come from calculus, and therefore will not be discussed in this course.

One of the methods used to compute π illustrates the approximation technique we mentioned earlier. You may omit this discussion if you find it too complicated, and just accept the value of π.

We draw a disk, and a grid consisting of vertical and horizontal lines:

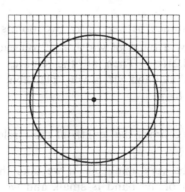

Figure 8.20

If the grid is fine enough, that is, if the sides of the square are suffi-
ciently small, then the area of the disk is approximately equal to the sum
of the areas of the squares which are contained in the disk. The difference between the area of the disk and this sum will be, at most, the sum
of the areas of the squares which intersect the circle. To determine the
area of the disk approximately, you just count all the squares that lie inside the circle, measure their sides, add up their areas, and get the desired approximation. Using fine graph paper, you can do this yourself
and arrive at your own approximation of the area of the disk. (See
Exercise 8.)

Of course, you want to estimate how good your approximation is.
The difference between the sum of the areas of all the little squares contained in the disk and the area of the disk itself is determined by all the
small portions of squares which touch the boundary of the disk, i.e.
which touch the circle. We have a very strong intuition that the sum of
such little squares will be quite small if our grid is fine enough, and in
fact, we give an estimate for this smallness in the following discussion.

Suppose that we make the grid so that the squares have sides of
length c. Then the diagonal of such a square has length $c\sqrt{2}$. If a
square intersects the circle, then any point on the square is at distance at
most $c\sqrt{2}$ from the circle. Look at Figure 8.21(a).

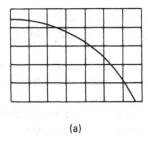   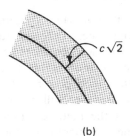

(a)   (b)

Figure 8.21

This is because the distance between any two points of the square is at
most $c\sqrt{2}$. Let us draw a band of width $c\sqrt{2}$ on each side of the circle,
as shown in Figure 8.21(b). Then all the squares which intersect the
circle must lie within that band. It is very plausible that the area of the
band is at most equal to

$$2c\sqrt{2} \text{ times the length of the circle.}$$

Thus if we take c to be very small, i.e. if we take the grid to be a very
fine grid, then the area of the band is small, and we see that the area of

the disk is approximated by the area covered by the squares lying entirely inside the disk.

This same type of argument also explains how areas of general regions can be approximated by a grid of squares.

The formula for the area of a disk may be applied to find the areas of a variety of regions. In doing computations, we may leave answers in terms of π, or we may use approximate values of π, such as $\frac{22}{7}$ or 3.14.

Example. Find the area of the band between concentric circles of radius 4 and 6:

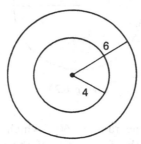

Figure 8.22

We have:

$$\text{Area of the disk of radius 6 is: } 6^2\pi = 36\pi.$$
$$\text{Area of the disk of radius 4 is: } 4^2\pi = 16\pi.$$

Subtracting gives us the area of the band, $36\pi - 16\pi = 20\pi$, in terms of π.

Example. A circle is inscribed in a square of area 36. What is the area enclosed by the circle?

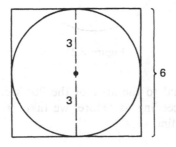

Figure 8.23

Since the area of the square is 36, a side has length 6. Therefore the radius of the circle is 3, and the area enclosed by the circle is $3^2\pi = 9\pi$. Using 3.14 for π, we get the approximate value 28.26 for the area enclosed by the circle.

Example. Find the area of the shaded region in the disk:

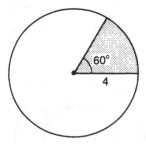

Figure 8.24

Since the angle between the radii is 60°, which is $\frac{1}{6}$th of the full angle 360°, the shaded area is $\frac{1}{6}$th of the area of the whole disk. The area of the disk is $4^2\pi = 16\pi$; therefore the shaded area is $16\pi/6$.

The region in the example is a special case of a sector. In general, a **sector** is the region of a disc contained in an angle.

Example. Find the area of the shaded region below:

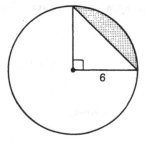

Figure 8.25

The shaded area is equal to the area of the 90° sector minus the area of the right triangle. Proceeding as before, we take $\frac{1}{4}$th the area of the disk (90° is $\frac{1}{4}$th of 360°) to find that

$$\text{area of the sector} = \tfrac{1}{4}\pi 6^2 = 9\pi.$$

The area of the triangle is $\frac{1}{2} \cdot 6^2 = 18$. Therefore the area of the desired region is $9\pi - 18$.

Example. Find the area of a sector having angle $31°$ in a disk of radius 5.

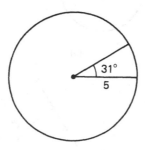

Figure 8.26

We have:
$$\text{Area of disk} = 5^2\pi = 25\pi.$$

The area of the sector is equal to the fraction $31/360$ of the total area. Hence
$$\text{Area of sector} = \frac{31}{360} \, 25\pi.$$

This is a perfectly good answer, and you don't need to simplify.

On the other hand, if you have a small calculator available, then it will be easy to convert this answer into an approximate decimal form.

The examples illustrate a general formula which relates the area of a sector and the measure of the angle, as follows.

Let A be an angle with vertex at a point P and let S be the sector determined by A in the disk D centered at P. Let A have x degrees, that is $m(A) = x°$. Then

$$x = \frac{\text{area of } S}{\text{area of } D} \, 360.$$

See Figure 8.27(a).

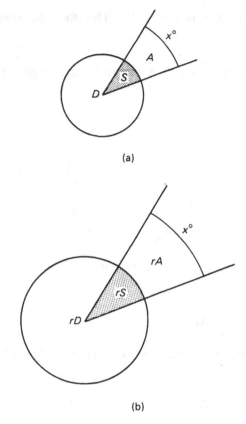

(a)

(b)

Figure 8.27

In the previous examples, we were given the measure of the angle x, the area of the disk D, and we found the area of the sector. On the other hand, given any two of these three quantities we can find the third one.

Example. Let D be a disk having area 54 cm^2, and let S be a sector in D having area 12 cm^2. Find the angle of the sector.

To do this, we use the formula, and get:

$$x = \frac{12}{54}\, 360 = \frac{2}{9}\, 360 = 2 \cdot 40 = 80.$$

Thus the answer is $80°$.

Example. Let D be a disk whose radius is 2 cm. Let S be a sector in D having area 7 cm^2. Find the angle of the sector.

The area of the disk is equal to $2^2\pi = 4\pi$. By the formula, we obtain

$$x = \frac{7}{4\pi}\,360 = \frac{630}{\pi}.$$

Thus the answer is $630/\pi$ degrees.

Observe that in the formula we have not specified the radius of the disk D. This is because the ratio

$$\frac{\text{area of } S}{\text{area of } D}$$

does not depend on the radius. To see this, suppose we perform a dilation by r. Then

$$\frac{\text{area of } rS}{\text{area of } rD} = \frac{r^2 \cdot \text{area of } S}{r^2 \cdot \text{area of } D} = \frac{\text{area of } S}{\text{area of } D} \qquad \text{(Figure 8.27(b))}$$

by canceling the factor r^2. This also shows:

Theorem 8-2. *The measure of an angle does not change under dilation, in other words, for any angle A,*

$$measure\ of\ rA = measure\ of\ A.$$

8, §2. EXERCISES

1. A rectangle has sides of length 30 mm and 60 mm. If the sides of the rectangle are tripled in length, what is the area of the larger rectangle?

2. The area of a square is 10 sq. cm. If the sides of the square are doubled in length, what is the area of the new square?

3. When the sides of any square are doubled in length, the area is multiplied by _____?

4. If the radius of a disk is doubled, its area is multiplied by _____?

5. Suppose you own a table which is a square with 1.5 m sides. You wish to build a second table whose area is twice as large as the first. How long must each side be?

6. If you wish to double the area of a square, you must multiply the sides by _____?

7. If you were an officer in the state consumer fraud commission, what would you point out to the company that put up the sign illustrated below?

> LOTS: 40 m × 100 m
> $2,000
>
> HALF-SIZE LOTS: 20 m × 50 m
> at less than half-price
> $900

8. Obtain a piece of fine grid graph paper (5 divisions to the centimeter will work well). Locate a point O at the center of the paper. Using a compass, draw a circle of radius 30 units with center O. Count the number of boxes contained within the circle (you will probably discover some efficient ways of doing this). If we say that each box has area of 1 square unit, your count will approximate the area of a disk with radius 30 units. Since

$$\text{area} = \pi r^2$$
$$= \pi 30^2$$
$$= \pi \cdot 900,$$

dividing your count by 900 should give a fairly good estimate of π.

9. Trace the border of the lake illustrated below on some graph paper. Estimate its area using a technique like that described in the text. Note the scale factor of 1 cm to the kilometer.

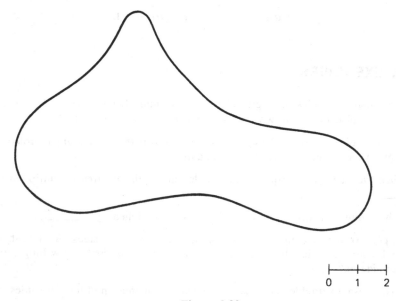

```
|———+———|
0   1   2
```

Figure 8.28

10. Find the area of the shaded regions:

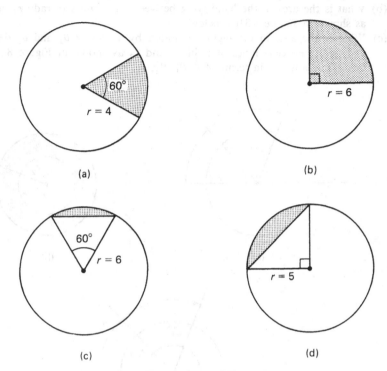

(a)   (b)

(c)   (d)

Figure 8.29

11. Let S be the shaded sector in the disk of radius 1 pictured below:

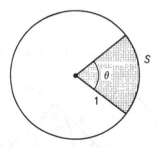

Figure 8.30

What is the area of the sector if the indicated angle θ has value:
(a) 90°   (b) 120°   (c) 30°   (d) $x°$
Express your answer in terms of π.

12. (a) What is the area of a sector in the disk of radius r lying between angles of θ_1 and θ_2 degrees, as shown in Figure 8.31(a) below?

(b) What is the area in the band lying between two circles of radii r_1 and r_2 as shown in Figure 8.31(b) below?

(c) What is the area in the region bounded by angles of θ_1 and θ_2 degrees and lying between circles of radii r_1 and r_2 as shown in Figure 8.31(c)? Give your answers in terms of π, θ_1, θ_2, r_1, r_2.

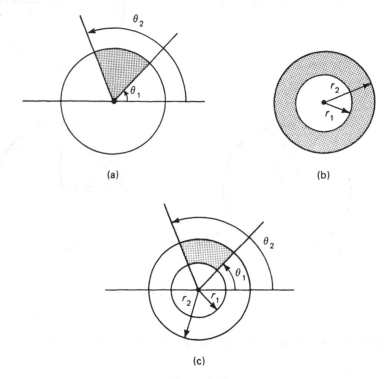

(a)   (b)

(c)

Figure 8.31

13. Two circles with radius 6 intersect in two points P and Q as shown:

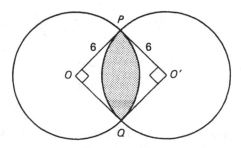

Figure 8.32

$m(\angle O) = m(\angle O') = 90°$ as shown. Compute the area of the shaded region. [*Hint*: Draw segment $\overline{PQ}$.]

14. Find the area of the shaded region: The curved lines are semicircles:

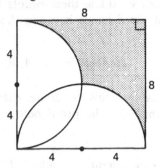

Figure 8.33

15. A large metal washer has an inner diameter of 6 cm and an outer diameter of 10 cm, as shown. Find the area of the face of the washer.

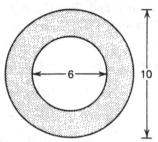

Figure 8.34

16. A regular octagon has been inscribed in a circle of radius r. Prove that the area of the regular octagon can be expressed as:

$$2 \cdot \sqrt{2} \cdot r^2.$$

[*Hint*: Draw the altitude $\overline{BX}$ as shown. What is $m(\angle BOA)$?]

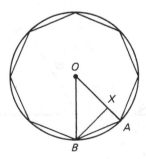

Figure 8.35

8, §3. CHANGE OF LENGTH UNDER DILATION

Let A and B be two points in the plane. What happens to the distance between A and B when we dilate these points by a positive number r?

The answer is that the distance between the points is multiplied by r; in other words,

$$d(rA, rB) = r \cdot d(A, B).$$

You have already justified this to some extent using the Pythagoras theorem in Experiment 8-1. The proof below using coordinates is similar.

Theorem 8-3. *Let r be a positive number. If A, B are points, then*

$$d(rA, rB) = r \cdot d(A, B).$$

Proof. Let $A = (a_1, a_2)$ and $B = (b_1, b_2)$. Then $rA = (ra_1, ra_2)$ and $rB = (rb_1, rb_2)$. Furthermore:

$$
\begin{aligned}
d(rA, rB) &= \sqrt{(rb_1 - ra_1)^2 + (rb_2 - ra_2)^2} \\
&= \sqrt{(r(b_1 - a_1))^2 + (r(b_2 - a_2))^2} \\
&= \sqrt{r^2(b_1 - a_1)^2 + r^2(b_2 - a_2)^2} \\
&= \sqrt{r^2[(b_1 - a_1)^2 + (b_2 - a_2)^2]} \\
&= r\sqrt{(b_1 - a_1)^2 + (b_2 - a_2)^2} \\
&= r \cdot d(A, B).
\end{aligned}
$$

Theorem 8-3 tells us what happens to the perimeters of geometric figures when we dilate them. Consider a triangle with sides of length a, b, and c (Figure 8.36(a)). The perimeter of this triangle is $a + b + c$.

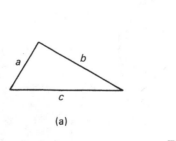

(a)

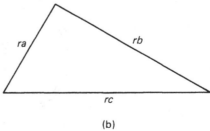

(b)

Figure 8.36

If we dilate the triangle by r, Theorem 8-3 tells us that the length of each side is multiplied by r (Figure 8.36(b)). The perimeter of the dilated triangle is

$$ra + rb + rc = r(a + b + c),$$

which is r times the perimeter of the original triangle. As we learned in the previous section, the *area* of the triangle is multiplied by r^2 when we dilate by r. Here we see that the *perimeter* changes by a factor of r.

What about a figure which does not consist of line segments? We can investigate what happens to the perimeter of a general region S in the plane under dilation by a number r using an approximation technique again.

Figure 8.37

The borders of such regions are curves. If C is any curve in the plane, we can approximate it by a series of line segments. Figure 8.38(a) shows how we can approximate a curve using 6 segments.

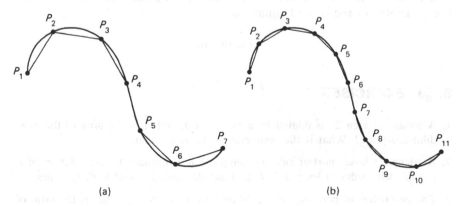

(a)   (b)

Figure 8.38

If we take more, smaller segments, we get a better approximation (Figure 8.38(b)). Note that this argument resembles the one we used concerning areas in the previous section. When we dilate curve C by a number r, each segment is also dilated by r. Since the length of each segment is multiplied by r, and the sum of the lengths approximates the length of the dilated curve, we see that the length of the dilated curve is multiplied by r.

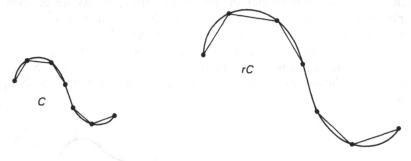

Figure 8.39

Thus it is plausible to conclude that the length of the border of an arbitrary region is multiplied by r when the region is dilated by r. However, we recall from the previous section that the *area* is multiplied by r^2 in the same situation.

Example. A square having area $36\,\text{cm}^2$ is dilated by a factor of 4. What is the area of the dilated square? What is the perimeter of the dilated square?

Under dilation by a factor of 4, area changes by a factor of $4^2 = 16$. Hence the area of the dilated square is $36 \cdot 16 = 576\,\text{cm}^2$.

On the other hand, the side of the square is $6\,\text{cm}$ because $6^2 = 36$. The perimeter is $6 \cdot 4 = 24\,\text{cm}$. Length changes by the factor of 4. Hence the perimeter of the dilated square is

$$4 \cdot 24 = 96\,\text{cm}.$$

8, §3. EXERCISES

1. A square of area 25 is dilated by a factor of 3. What is the area of the new, dilated, square? What is the perimeter of the new square?

2. A triangle whose shortest side has length 7 is congruent to the dilation of a triangle with sides of lengths 2, 4, 5. Find the perimeters of both triangles.

3. The perimeters of two squares are 10 and 16 respectively. What is the ratio of their areas?

4. The perimeters of two squares are a and b respectively. What is the ratio of their areas?

8, §4. THE CIRCUMFERENCE OF A CIRCLE

The **circumference** of a circle is defined as the length of the circle. In this section, we prove the standard formula for the circumference c of a circle of radius r:

$$c = 2\pi r.$$

Proof. Let D be a disk of radius r. We divide D into n sectors, whose angles have

$$\frac{360}{n} \text{ degrees.}$$

Here, n is an integer. The picture is that of Figure 8.40, drawn with $n = 7$.

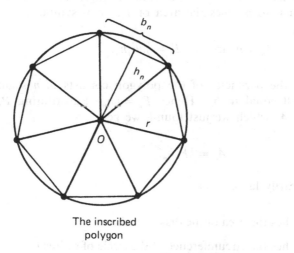

The inscribed
polygon

Triangle T_n

Figure 8.40

You should now do Exercise 1, taking special values for n, namely $n = 4$, $n = 5$, $n = 6$, $n = 8$ (more if you wish). In each case, determine the number of degrees

$$\frac{360}{4}, \quad \frac{360}{5}, \quad \frac{360}{6}, \quad \frac{360}{8}, \quad \text{etc.}$$

and draw the corresponding sectors. In general, we then join the end-points of the sectors by line segments, thus obtaining a polygon inscribed in the circle. The region bounded by this polygon consists of n triangles congruent to the same triangle T_n, lying between the segments from the origin O to the vertices of the polygon. Thus in the case of 4 sides, we call the triangle T_4. In the figure with 5 sides, we call the triangle T_5. In the figure with 6 sides, we call the triangle T_6. In the figure with 8 sides, we call the triangle T_8. And so on, to the polygon having n sides, when we call the triangle T_n.

We denote the (length of the) base of T_4 by b_4 and its height by h_4. We denote the base of T_5 by b_5 and its height by h_5. In general, we denote the base of T_n by b_n and its height by h_n, as indicated on Figure 8.40. Since the area of a triangle whose base has length b and whose height has length h is $\frac{1}{2}bh$, we see that the area of our triangle T_n is given by

$$\text{area of } T_n = \tfrac{1}{2}b_n h_n.$$

Let A_n be the area of the region surrounded by the polygon, and let P_n be the perimeter of the polygon. Since the polygonal region consists of n triangular regions congruent to the same triangle T_n, we find that the area of A_n is equal to n times the area of T_n, or in symbols,

$$A_n = n \cdot \text{area of } T_n = \tfrac{1}{2}nb_n h_n.$$

On the other hand, the perimeter of the polygon consists of n segments whose lengths are all equal to b_n. Hence $P_n = nb_n$. Substituting P_n for nb_n in the value for A_n which we just found, we get

$$A_n = \tfrac{1}{2}P_n h_n.$$

As n becomes arbitrarily large,

A_n approaches the area of the disk of radius r,

P_n approaches the circumference of the circle of radius r,

and

h_n approaches the radius r of the disk.

For instance, if we double the number of sides of the polygon successively, the picture looks like Figure 8.41.

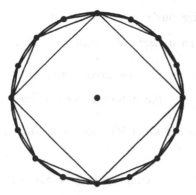

Figure 8.41

Let c denote the circumference of the circle of radius r. Then A_n approaches πr^2. Since $A_n = \frac{1}{2}P_n h_n$, it follows that A_n also approaches $\frac{1}{2}cr$. Thus we obtain

$$\pi r^2 = \tfrac{1}{2}cr.$$

We cancel r from each side of this equation, and multiply both sides by 2. We conclude that

$$c = 2\pi r,$$

as was to be shown.

As a special case, we see that the circumference of the circle of radius 1 has length 2π.

That the circumference of the circle of radius r is r times 2π is consistent with our earlier observation that the length of a curve is multiplied by r under dilation by r.

Remark. In Chapter 9, §5 you will find another proof for the formula $c = 2\pi r$, based on an entirely different principle, assuming only the formula for the area. You can read that proof right away if you like. The reason why this other proof is postponed to the later chapter is that the idea for that proof will be used to determine the area of a sphere. Such interconnections between various topics show why it is often valuable not to read a book in the order in which it is written.

We now summarize our results about circles in

Theorem 8-4. *Given a circle with radius r. Then:*

$$the \ area = \pi r^2,$$

$$the \ circumference \ c = 2\pi r.$$

Example. What is the circumference of a circle of radius 5? Circumference $= 2\pi 5 = 10\pi$.

Example. The area of a circle is 36π. What is its circumference?

$$Area = \pi r^2 = 36\pi.$$

Thus $r = 6$, and the circumference is

$$2\pi \cdot 6 = 12\pi.$$

Example. Suppose that a disk has an area of 23π cm². What is the length of the surrounding circle?

Let r be the radius of the disk. We know that

$$23\pi = \pi r^2.$$

Hence $r^2 = 23$ and

$$r = \sqrt{23}.$$

Therefore the length of the circle is $2\pi r$, or in other words, $2\pi\sqrt{23}$. Since the area is given as cm² units, the length turns out in cm, so the length of the circle is

$$2\pi\sqrt{23} \text{ cm.}$$

Example. Four tin cans of radius 12 cm have to be tied together with a plastic band as shown on the figure. How long must the band be?

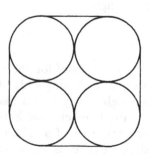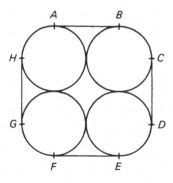

Figure 8.42

The band consists of four segments which are horizontal or vertical, and four curved pieces at the four corners. Label points as shown on the figure. The segments are then

$$\overline{AB}, \quad \overline{CD}, \quad \overline{EF}, \quad \overline{GH}.$$

Each segment has length equal to twice the radius of the cans, so each segment has length 24 cm. Hence the four segments together have length $4 \cdot 24 = 96$ cm.

The curved parts of the band go over the cans, and each curved part covers $\frac{1}{4}$th of the circumference, as shown on the next figure, for one can.

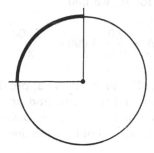

Figure 8.43

Each such curved part has length equal to

$$\frac{1}{4} 2\pi r = \frac{1}{4} 2\pi \cdot 12 = 6\pi \text{ cm.}$$

Hence the four curved parts have total length $= 4 \cdot 6\pi = 24\pi$ cm.

The total length of the band is therefore equal to the sum of the four segments and the four curved parts, and is equal to

$$96 + 24\pi \text{ cm.}$$

This is a correct answer. If you want an approximation you can use the approximate value $\pi = 3.14$ to get a decimal answer but it is just as well to leave the answer in the other form until you actually need to cut the band.

Example. A car is going 60 km/hr. Each wheel has a radius of 0.4 m. How many revolutions per hour is the wheel turning?

Let $N =$ number of revolutions per hour. Let c be the circumference of the wheel. Let S be the number of kilometers traveled per hour. Then we have the formula

$$S = Nc.$$

Therefore

$$N = \frac{S}{c}.$$

We are given $S = 60$. The radius has $0.4/1,000$ km because 1 km = 1,000 m. Hence the circumference is

$$c = 2\pi r = \frac{0.8\pi}{1,000} \text{ km}.$$

Plugging in the formula for N, we find

$$N = \frac{60}{0.8\pi} \, 1,000 = \frac{75,000}{\pi}.$$

This is a correct answer. If you want an approximate decimal, you can use $\pi = 3.14$ to 2 decimals, and then you find approximately 23,885 revolutions per hour. Unless you have great need for it, it is just as well to leave the answer with π in it, without converting to decimals.

8, §4. EXERCISES

1. Draw the picture of a polygon with sides of equal length, inscribed in a circle of radius 5 cm, in the cases when the polygon has:
   (a) 4 sides,   (b) 5 sides,   (c) 6 sides   (d) 8 sides,   (e) 9 sides.
   Draw the radii from the center of the circle to the vertices of the polygon. Use a protractor for the angles of the sectors. Using a ruler, measure (approximately) the base of each triangle and measure its height. From your measurements, compute the area inside the polygon, and the circumference of the polygon. Compare these values with the area of the disk and its circumference, given as πr^2 and $2\pi r$ respectively, and $r = 5$. Use the approximate value $\pi = 3.14$.

2. Get a tin can with as big a circular bottom as possible. Take a tape, measure the circumference of the bottom, measure the diameter, take the ratio and get a value for π, probably good to one decimal place. Do the same thing to another circular object, say a frying pan, and verify that you get the same value for π.

3. If a disk has an area of (a) 24π cm^2, (b) 36π cm^2, what is the length of the surrounding circle?

4. If a circle has a length of (a) 15π cm, (b) 9π cm, (c) 20π cm, what is the area of the enclosed disk?

5. A rectangular sheet of metal is bent to form a cylinder. The radius of the base is 3 m, and the height of the cylinder is 5 m. What is the total surface area of the cylinder?

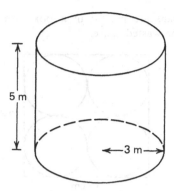

Figure 8.44

6. A rectangular sheet of metal is bent to form half a cylinder, as shown. If the radius of the base of the cylinder is 4 m, and the height of the cylinder is 11 m, what is the total area of the sheet of metal?

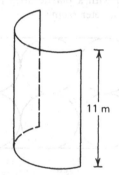

Figure 8.45

7. Suppose the roof of a cabin is in the shape of half a cylinder as shown.
   (a) What is the surface area of the roof?
   (b) What is the total area of the roof and the sides of the cabin?

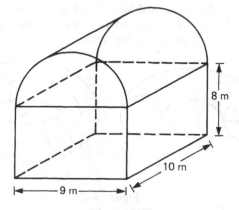

Figure 8.46

8. Four equal sized cans are packed in a box with a square base as shown. Find the area of the wasted space.

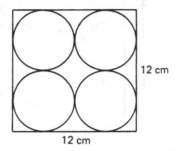

12 cm

12 cm

Figure 8.47

9. If the radius of a circle is doubled, what happens to the circumference?

10. Six beer cans are wrapped with a plastic strip as shown. How long must the strip be if each can has diameter 6 cm?

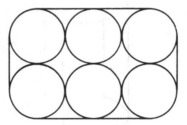

Figure 8.48

11. A bicycle is going 15 km/hr. A wheel has diameter equal to 1.2 m. How many revolutions per hour is the wheel turning?

12. Derive the formula which gives the area of a disk in terms of its perimeter.

13. Visualize a perfect doughnut as shown. Let b be the radius of a cross section of the doughnut, and let a be the radius of a circle passing through the middle of the doughnut, as shown on the figure.

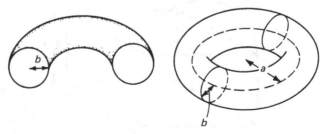

Figure 8.49

(a) Guess what the total volume of the doughnut should be.

(b) Guess what the total surface area of the doughnut should be.

Give the answers in terms of a, b, π. The technical name for a doughnut is a **torus**.

14. Suppose that the circumference of the earth is 40,000,000 m along a great circle. Assume that the earth is a perfect sphere, and that a plastic band has been tightly fit around the earth along a great circle. Now suppose we cut the band, and add a piece to it which is 1 m long. We then place the band again like a circular belt around the earth. How good is the fit? Do you think there is barely room for a piece of paper to slide between the earth and this belt? Do you think there is enough room for a person to walk upright in between? After you have thought about this, why don't you use the formula for the circumference of a circle and *prove* what the answer actually is. How relevant was the fact that a great circle is 40,000,000 m long?

ADDITIONAL EXERCISES ON CIRCLES

1. A regular octagon is inscribed inside a circle of radius x. Prove that the length of a side of the regular octagon can be expressed as $x\sqrt{2 - \sqrt{2}}$.

2. A regular 10-gon inscribed in a circle of radius 1 has a side of length $\frac{1}{2}(\sqrt{5} - 1)$. Describe how one might construct a line segment with this length. [*Hint*: Using Pythagoras, first construct a segment of length $\sqrt{5}$.]

3. In the picture below, the two circles are externally tangent at point O; $\overline{BC}$ is tangent to both circles; the radius of the large circle is 25 and the radius of the small circle is 9. Find $|BC|$.

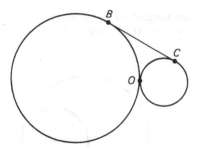

Figure 8.50

4. Two equilateral triangles are circumscribed about the same circle such that their intersection is a regular hexagon (see picture below). If the radius of the circle is 10, find values for the following:

(a) The area and the perimeter of the hexagon.

(b) The area and the perimeter of the six pointed star formed by the union of the two equilateral triangles.

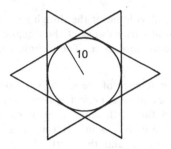

Figure 8.51

5. If the circumference of a circle is 16π, what is the perimeter of a regular hexagon inscribed in it?

6. A square is inscribed in a circle. If the circumference of the circle is 80π cm, find the length of a side of the square.

7. A wheel rolls along the ground at a speed of 350 revolutions per minute. If the diameter of the wheel is 20 cm, find the ground speed of the wheel in cm per minute.

8. If a square is circumscribed about a circle whose circumference is 10π, find the perimeter of the square.

9. In a circle whose radius is 6 cm, find the number of degrees in the central angle of an arc whose length is 4π cm.

10. In a circle, an arc of $72°$ has a length of 4π. Find the length of the radius.

11. If the circumference of a circle is 10π, find its area.

12. An 8 by 12 rectangle is inscribed in a circle. Find the area of the circle.

13. Circles A and B are tangent to each other internally, and circle B passes through the center of A. If the area of circle A is 16, find the area of circle B.

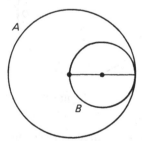

Figure 8.52

14. Quadrilateral $PQRS$ is inscribed in a circle of radius 10. If $m(\angle PQR) = 150°$, find the length of arc $\overgroup{PQR}$.

8, §5. SIMILAR TRIANGLES

At the beginning of this chapter, we illustrated two quadrilaterals which were not congruent, but which were related in some way. We can now accurately define this relationship.

We shall say that two figures in the plane are **similar** whenever one is congruent to a dilation of the other. Therefore the two quadrilaterals mentioned above are similar, since one is just an enlargement of the other. Any two circles are similar. If the two circles have the same radius, we simply take dilation by 1 to satisfy the definition.

For the moment, we will study similar triangles, as illustrated in Figure 8.53.

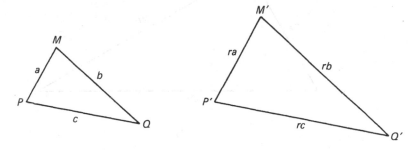

Figure 8.53

We can easily generate similar triangles by dilating a triangle with respect to one of its vertices (Figure 8.54(a)), or with respect to a point O not a vertex (Figure 8.54(b)).

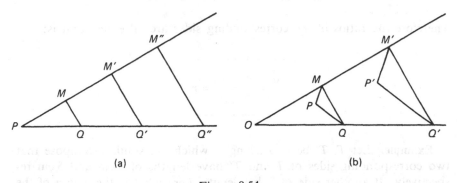

(a)   (b)

Figure 8.54

Let T be a triangle whose sides have lengths a, b, c respectively. If we dilate T by a factor of r, we obtain a triangle which we denote by rT.

The lengths of its sides will be ra, rb, rc, as we saw in the preceding section. Note that r can be any positive number. For instance in Figure 8.55(a), (b), (c), we have drawn triangles T, $\frac{1}{2}T$, and $2T$.

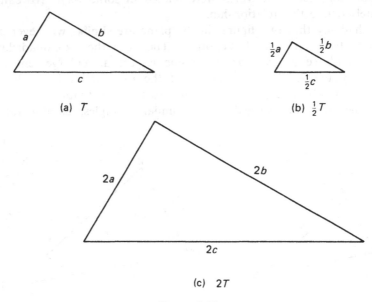

(a) T (b) $\frac{1}{2}T$

(c) $2T$

Figure 8.55

Denote by T' the dilation of T by r. Let a', b', c' be the lengths of the corresponding sides. Then we have

$$a' = ra, \qquad b' = rb, \qquad c' = rc.$$

Therefore the ratios of the corresponding sides are all equal, that is:

$$\frac{a'}{a} = \frac{b'}{b} = \frac{c'}{c} = r.$$

Example. Let T, T' be two triangles which are similar. Suppose that two corresponding sides of T and T' have lengths of 3 cm and 5 cm respectively. If another side of T has length 7 cm what is the length of the corresponding side of T'?

To do this, we use the previous formula. Let

$$a = 3, \qquad a' = 5, \qquad b = 7.$$

We have to find b'. The formula tells us that

$$\frac{5}{3} = \frac{b'}{7}.$$

Hence

$$b' = \frac{7 \cdot 5}{3} = \frac{35}{3}.$$

The desired length is therefore 35/3 cm.

Example. Let T be the triangle $\triangle ABC$, as shown on the figure.

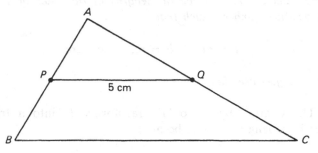

Figure 8.56

Let P be the midpoint of $\overline{AB}$ and Q the midpoint of $\overline{AC}$. If $\overline{PQ}$ has length 5 cm, find the length of $\overline{BC}$.

To do this, we observe that $\triangle ABC$ is the dilation by 2 of $\triangle APQ$, and $\overline{BC}$ is the side corresponding to $\overline{PQ}$. Hence

$$|BC| = 2|PQ| = 2 \cdot 5 = 10 \text{ cm}.$$

Example. Again let T be the triangle $\triangle ABC$ as shown in the figure, but let P be the point one third of the distance from A to B, and let Q be the point one-third of the distance from A to C.

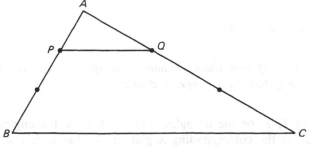

Figure 8.57

If $\overline{PQ}$ has length 7 cm, find the length of $\overline{BC}$.

This time, we note that $\triangle ABC$ is the dilation by 3 of $\triangle APQ$, and $\overline{BC}$ is the side corresponding to $\overline{PQ}$. Hence

$$|BC| = 3|PQ| = 3 \cdot 7 = 21 \text{ cm.}$$

We have seen that if two triangles are similar, then the ratios of the lengths of corresponding sides are equal to a constant r. We now prove the converse.

Theorem 8-5. *Let T, T' be triangles. Let a, b, c be the lengths of the sides of T, and let a', b', c' be the lengths of the sides of T'. If there exists a positive number r such that*

$$a' = ra, \qquad b' = rb, \qquad c' = rc,$$

then the triangles are similar.

Proof. The dilation by $1/r$ of T' transforms T' into a triangle T'' whose sides have lengths a, b, c, because

$$\frac{1}{r} \cdot ra = a, \qquad \frac{1}{r} \cdot rb = b, \qquad \frac{1}{r} \cdot rc = c.$$

Therefore T, T'' have corresponding sides of the same length. By condition **SSS** we conclude that T and T'' are congruent. Therefore T is congruent to a dilation of T', and triangles T and T' are similar.

From Theorem 8-2 we know that:

Theorem 8-6. *If two triangles are similar, then their corresponding angles have the same measure.*

The converse is also true.

Theorem 8-7. *If two triangles have corresponding angles having the same measure, then the triangles are similar.*

Proof. Let T, T' be the triangles. Let A, B, C be the angles of T and let A', B', C' be the corresponding angles of T'. Let a, b, c and a', b', c' be the lengths of corresponding sides. Let

$$r = a'/a$$

be the ratio of the lengths of one pair of corresponding sides. Then $a' = ra$. Dilation by r transforms T into a triangle T'' whose sides have lengths

$$a'' = ra, \qquad b'' = rb, \qquad c'' = rc$$

respectively. The triangles T' and T'' have one corresponding side having the same length, namely

$$a'' = a' = ra.$$

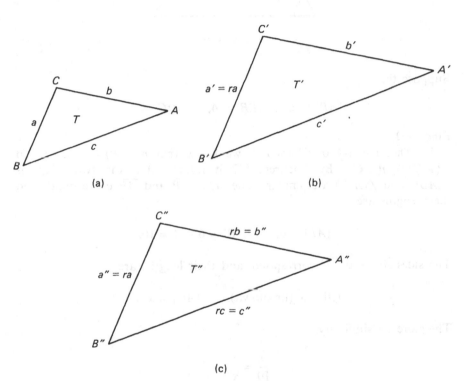

(a)

(b)

(c)

Figure 8.58

We have seen in the previous theorem that dilation preserves the measures of angles. Hence the angles adjacent to this side in T' and T'' have the same measure, that is:

$$m(\angle B') = m(\angle B'') \qquad \text{and} \qquad m(\angle C') = m(\angle C'').$$

It follows from the **ASA** property that T', T'' are congruent. Hence T' is congruent to a dilation of T, and hence T' is similar to T, as was to be shown.

Example. Given $\triangle ABC$, let P be a point on $\overline{AB}$, and let Q be a point on $\overline{AC}$ as shown on the figure. Assume that $\overline{PQ}$ is parallel to $\overline{BC}$.

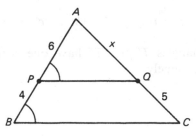

Figure 8.59

Suppose that

$$|AP| = 6, \qquad |PB| = 4, \qquad |QC| = 5.$$

Find $|AQ|$.

By Theorem 1-4 of Chapter 1, we know that $m(\angle P) = m(\angle B)$ and $m(\angle Q) = m(\angle C)$. By Theorem 8-7 it follows that the two triangles $\triangle ABC$ and $\triangle APQ$ are similar. The sides $\overline{AP}$ and $\overline{AB}$ correspond, and their lengths are

$$|AP| = 6, \qquad |AB| = 6 + 4 = 10.$$

The sides $\overline{AQ}$ and $\overline{AC}$ correspond, and their lengths are

$$|AQ| = x \text{ (unknown)}, \qquad |AC| = x + 5.$$

Therefore by similarity,

$$\frac{6}{10} = \frac{x}{x + 5}.$$

We must solve for x. Cross multiplying yields

$$6x + 30 = 10x.$$

Therefore

$$4x = 30, \qquad x = \frac{30}{4} = \frac{15}{2}.$$

The answer is $|AQ| = 15/2$.

In the above example, many people feel that one should obtain the answer also from the equation

$$\frac{6}{4} = \frac{x}{5},$$

and then solve for x. It is true that this gives the right answer, but it does not follow from the general similarity formula which we have given without some additional algebra. We shall now state and prove the theorem which justifies this.

Theorem 8-8. *Given triangle* $\triangle ABC$, *let P be a point on* $\overline{AB}$, *let Q be a point on* $\overline{AC}$, *and assume that* $\overline{PQ}$ *is parallel to* $\overline{BC}$. *Then*

$$\frac{|AP|}{|PB|} = \frac{|AQ|}{|QC|}.$$

Proof. Let $x = |AP|$, $y = |PB|$, $u = |AQ|$ and $v = |QC|$ as shown on the figure.

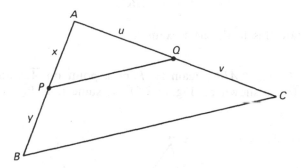

Figure 8.60

By Theorem 1-4 of Chapter 1 and Theorem 8-7 we know that $\triangle APQ$ and $\triangle ABC$ are similar. Hence

$$\frac{x}{x+y} = \frac{u}{u+v}.$$

Cross multiplying yields

$$xu + xv = xu + yu.$$

We can cancel xu to obtain

$$xv = yu.$$

We divide both sides by v, and also by y to find

$$\frac{x}{y} = \frac{u}{v},$$

as was to be shown.

Warning. Some people might feel that in the situation of Theorem 8-8, we should have the relation

$$\frac{|AP|}{|PB|} = \frac{|PQ|}{|BC|}.$$

These people are wrong! The correct relation is

$$\frac{x}{x + y} = \frac{|PQ|}{|BC|}.$$

We illustrate this in the next example.

Example. Given $\triangle ABC$, again let P be a point on $\overline{AB}$, and let Q be a point on $\overline{AC}$ as shown on Figure 8.61. Assume that $\overline{PQ}$ is parallel to $\overline{BC}$.

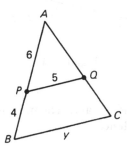

Figure 8.61

Suppose that

$$|AP| = 6, \qquad |PB| = 4, \qquad |PQ| = 5.$$

Find $|BC|$.

By Theorem 1-4 of Chapter 1, we know that $m(\angle P) = m(\angle B)$ and $m(\angle Q) = m(\angle C)$. By Theorem 8-7 it follows that the two triangles $\triangle ABC$ and $\triangle APQ$ are similar. The sides $\overline{AP}$ and $\overline{AB}$ correspond, and their lengths are as before,

$$|AP| = 6, \qquad |AB| = 6 + 4 = 10.$$

Let $y = |BC|$ be the unknown. The sides $\overline{PQ}$ and $\overline{BC}$ correspond, and their lengths are

$$|BC| = y, \qquad |PQ| = 5.$$

Therefore by similarity,

$$\frac{6}{10} = \frac{5}{y}.$$

We solve for y and obtain

$$y = \frac{5 \cdot 10}{6} = \frac{50}{6}.$$

This is a correct answer. It can be simplified a little to give the equally correct answer

$$\frac{25}{3}.$$

The next theorem gives an important example of similar triangles occurring frequently in geometry.

Theorem 8-9. *Let $\triangle ABC$ be a right triangle with right angle at A. Let $\overline{AD}$ be the perpendicular segment from A to $\overline{BC}$. Then the triangles*

$$\triangle ABC, \quad \triangle ABD, \quad \triangle ADC$$

are all similar.

Proof. We shall prove that $\triangle ABC$ is similar to $\triangle ADC$, and we leave the other cases as exercises. By Theorem 8-7, it suffices to prove that corresponding angles of $\triangle ABC$ and $\triangle ADC$ have the same measure.

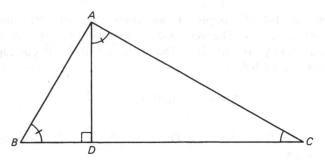

Figure 8.62

In fact:

>Angles $\angle BAC$ and $\angle ADC$ have the same measure 90°.
>The angle $\angle ACB$ is common to both triangles.

Since the sum of the angles of a triangle has 180°, it follows that the third angle $\angle CBA$ has the same measure as $\angle CAD$. This proves what we wanted.

Exercise (i) Prove that $\triangle ABC$ is similar to $\triangle DBA$.
(ii) Prove that $\triangle ABD$ is similar to $\triangle ACD$.

Do you have to go through the same type of argument again, or can you use a general principle at this point? Think about it for a moment, then use the next statement: If a figure U is similar to V and V is similar to W, then U is similar to W.

8, §5. EXERCISES

1. Let T and T' be similar triangles. Let a and a' be the lengths of corresponding sides. If T has a side of length 7 cm find the length of the corresponding side of T' whenever a, a' have the following values.
   (a) $a = 2$ cm and $a' = 4$ cm
   (b) $a = 3$ cm and $a' = 8$ cm
   (c) $a = 1$ cm and $a' = 5$ cm

2. In each of the triangles below, what dilation will map $\overline{AB}$ onto $\overline{A'B'}$, given that $\overline{AB} \| \overline{A'B'}$?

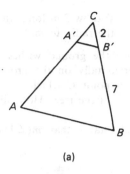 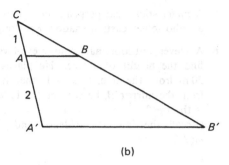

(a)   (b)

Figure 8.63

3. If $\overline{EF}\|\overline{GH}$ in the diagrams below, find $|DE|$ in each case and explain your answers.

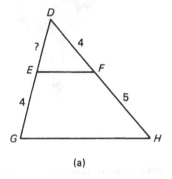   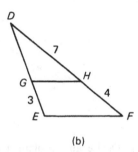

(a)   (b)

Figure 8.64

4. In triangle $\triangle ABC$, $m(\angle ADE) = m(\angle C)$. Prove that $\triangle ABC$ is similar to $\triangle ADE$. How long is $\overline{AB}$?

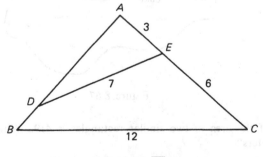

How long is $\overline{AB}$?

Figure 8.65

5. A meter stick held perpendicular to the ground has a shadow 2 m long when a radio tower casts a shadow 8 m long. How high is the radio tower?

6. A clever outdoorsman whose eye-level is 2 m above the ground wishes to find the height of a tree. He places a mirror horizontally on the ground 20 m from the tree, and finds that if he stands at a point C which is 2 m from the mirror B, he can see the reflection of the top of the tree. How high is the tree?
(Recall the law of incidence and reflection which states that $m(\angle 1) = m(\angle 2)$.)

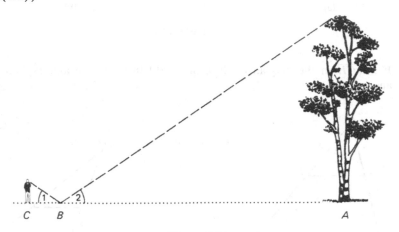

Figure 8.66

7. Devise and explain a method to determine, using similar triangles, the distance from point A to point B by measurements made only on the ground.

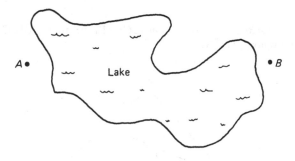

Figure 8.67

8. The ratio of the area of two similar rectangles is 4:9. What is the ratio of their perimeters?

9. The sides of a triangle are 5 cm, 7 cm, and 8 cm. Find the sides of a similar triangle if its longest side is 10 cm.

10. State a sufficient condition for two rectangles to be similar.

11. Two similar rectangles have lengths 18 m and 12 m. What is the ratio of their widths? What is the ratio of their perimeters?

12. The sides of a triangle are 5 m, 7 m, and 8 m long. Find the sides of a similar triangle if its shortest side is 10 m.

13. Let L_1, L_2, L_3 be three parallel lines as shown on the figure. Let K, K' be two lines intersecting L_1, L_2, L_3 as shown. Prove that $x/y = u/v$, where x, y, u, v are the lengths shown in the figure.

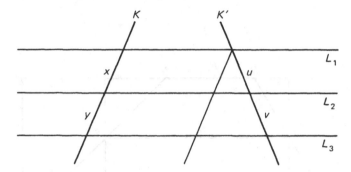

Figure 8.68

[*Hint*: Let P be the point of intersection of K' and L_1. Let K_1 be the line through P parallel to K. This should reduce the present problem to a theorem proved in the text.]

14. In the diagram, the lines L_1, L_2, L_3 are parallel. Find the length x in each case.

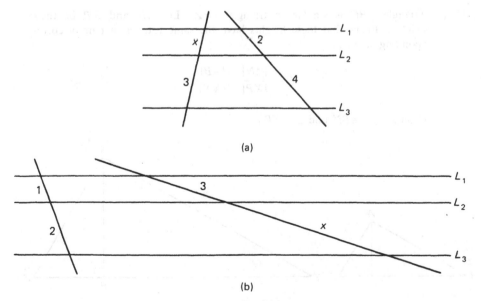

(a)

(b)

Figure 8.69

15. Prove: If two triangles are similar, and one pair of corresponding sides have the same length then the triangles are congruent.

16. Consider a trapezoid with vertices P, Q, M, N as shown with parallel sides $\overline{PQ}$ and $\overline{MN}$. Let R, S be the midpoints of $\overline{PM}$ and $\overline{QN}$ respectively. Let b be the length of $\overline{RS}$.
   (a) Prove that the area of a trapezoid is equal to bh, where h is the height, i.e. the perpendicular distance between $\overline{MN}$ and $\overline{PQ}$.
   (b) If b_1, b_2 are the lengths of $\overline{MN}$ and $\overline{PQ}$ respectively, prove that

$$b = \frac{b_1 + b_2}{2}.$$

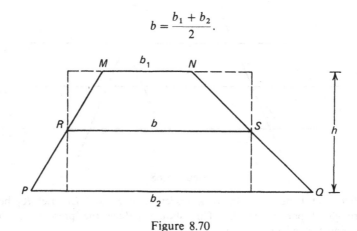

Figure 8.70

[*Hint*: For part (a), use triangles as shown by the dotted lines.]

17. (a) Triangle ABC is similar to triangle XYZ. Let AN and XP be their heights. Prove that these heights have the same ratio as a pair of corresponding sides, i.e.

$$\frac{|AN|}{|XP|} = \frac{|AB|}{|XY|}.$$

(Consider $\triangle ABN$ and $\triangle XYP$.)

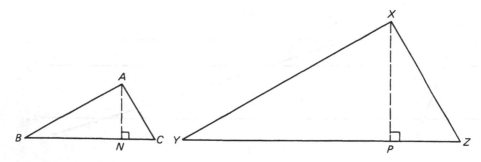

Figure 8.71

(b) Use part (a) to do this problem:

The bases of a trapezoid are 9 cm and 15 cm long, and the trapezoid has an altitude of 12 cm. If the diagonals intersect at *O*, find the distance of *O* from the longer base.

18. Given right triangle *ABC*, with $\overline{CN} \perp \overline{AB}$, and lengths as marked.

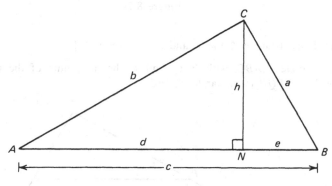

Figure 8.72

Show that (i) $b^2 = cd$, (ii) $a^2 = ce$, (iii) $h^2 = de$. [*Hint*: Use Theorem 8-9, write the appropriate ratios, and cross-multiply.]

19. Using the figure of Exercise 18, find the perimeter of $\triangle ABC$ if $e = 6$ and $h = 8$.

20. In right triangle *ABC*, let $\overline{BD} \perp \overline{AC}$, let $|BD| = 8$ and $|DC| = 4$. Find $|AD|$, $|BC|$, and $|AB|$.

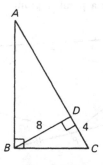

Figure 8.73

21. Suppose you build in your backyard a geodesic dome whose cross section is a semicircle, as illustrated. If the diameter of the dome is 5 m, how high would a wall be if erected 1 m from the edge?

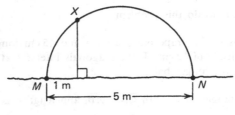

Figure 8.74

[*Hint*: Draw triangle $\triangle XMN$, and use Exercise 18.]

22. Given triangle $\triangle ABC$, with P, Q, and R the midpoints of the three sides. Prove that $\triangle PQR$ is similar to $\triangle ABC$.

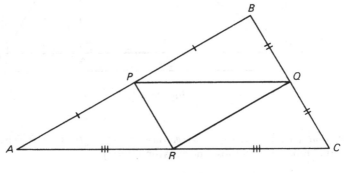

Figure 8.75

23. Prove that the line segment connecting the midpoints of two sides of a triangle is parallel to and one-half the length of the third side.

24. In quadrilateral MEAN, the points P, Q, R. S are midpoints of the four sides. Prove that $PQRS$ is a parallelogram.

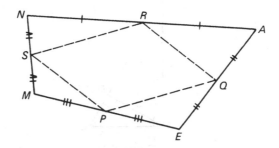

Figure 8.76

[*Hint*: Draw diagonal $\overline{MA}$, and use Exercise 23. Repeat with $\overline{EN}$.]

25. Use the results of Exercise 18 to prove the Pythagoras theorem.

CHAPTER 9

Volumes

You probably already have a good intuitive notion of volume, which measures the amount of space enclosed by a 3-dimensional figure. Just as we did for 2-dimensional figures to compute area, we define a unit volume to be the volume of a cube whose edges are one unit long. Therefore a cube with edges 1 cm long has volume 1 cubic centimeter, which we abbreviate 1 cm³. In this section we show how to compute the volumes of various figures.

9, §1. BOXES AND CYLINDERS

We start with the simplest types of volumes, those of rectangular boxes. If the sides of a rectangular box have lengths a, b, c then the volume of the box is given by the formula

$$V = abc.$$

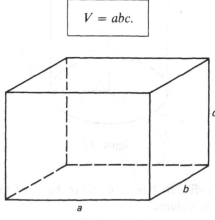

Figure 9.1

Example. If the sides of a box have lengths 3 cm, 4 cm, 7 cm then the volume of the box is

$$3 \cdot 4 \cdot 7 = 84 \, \text{cm}^3.$$

One can view a box as a 3-dimensional set of points lying above the base. For instance, in Figure 9.1 we may view the base of the box to be the rectangle with sides a, b, and we may view c as the height h of the box. The area of the base is

$$B = ab.$$

Then the volume of the box may be expressed as

$$\boxed{V = Bh,}$$

where B is the area of the base, and h is the height.

This same formula is valid for more general figures.

Example. Let D be a disk of radius r, and view D as the base of a cylindrical box of height h above D as shown on Figure 9.2. The area of D is $B = \pi r^2$. The volume of the cylinder is then

$$V = \pi r^2 h.$$

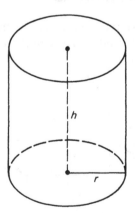

Figure 9.2

Example. If a cylinder has a circular base of diameter 6 cm and length 14 cm, find its volume.

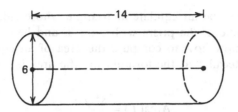

Figure 9.3

The radius of the base is 3 cm. Hence the volume is

$$V = \pi 3^2 \cdot 14 = 126\pi \text{ cm}^3.$$

We will now define the general notion of a cylinder with arbitrary base. Let S be a region in the plane, and let h be a positive number. Given a direction perpendicular to the plane, the set of points in 3-dimensional space which lie at distance $\leq h$ from a point of S in the given direction is called the **right cylinder with base S and height h.**

Thus an ordinary cylinder is a right cylinder with circular base. A rectangular box as considered in the first example is a right cylinder with rectangular base. Of course, we can make up right cylinders with other types of bases.

Example. Let T be a triangle, as in Figure 9.4(a), or a polygon, as in Figure 9.4(b). The right cylinder with base T and height h is called a **right prism.**

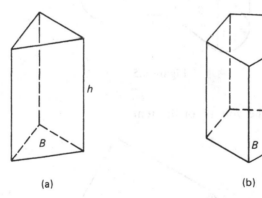

(a)   (b)

Figure 9.4

In each of these cases, the volume of the right cylinder (prism) satisfies the same formula as before,

$$V = Bh.$$

Example. Let T be an equilateral triangle whose sides have length 2. What is the volume of the prism with base T and height 7?

You should know how to compute the area of an equilateral triangle, given a side s. Recall that the formula for the area is

$$\text{Area}(T) = \frac{s^2\sqrt{3}}{4}.$$

In the present case, the area of the base is therefore $\sqrt{3}$. Consequently, the volume of the prism is equal to

$$V = Bh = \sqrt{3} \cdot 7 = 7\sqrt{3}.$$

In Figure 9.5 we have drawn a right cylinder with a fairly arbitrary base.

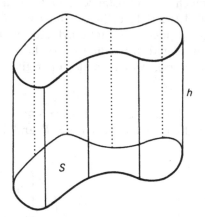

Figure 9.5

Example. Find the volume of the tent

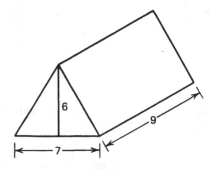

Figure 9.6

This tent is a right prism, whose base is a triangle T. The area of T is given by

$$B = \tfrac{1}{2} 7 \cdot 6 = 21.$$

Hence the volume of the tent is

$$V = 21 \cdot 9 = 189.$$

Some figures in space are obtained in a manner similar to the solids we have considered, but using some direction other than the perpendicular direction. Let S be a region in the plane. Let h be a given positive number. Suppose also a given direction. The set of points X which lie on a ray whose vertex is a point of S, in the given direction, and such that the **perpendicular distance** of X to the plane is $\leq h$, is called the **cylinder with base S and height h in the given direction**. We have drawn such a cylinder in Figure 9.7. We emphasize that h is the **perpendicular height**, not what is sometimes called the "slant height".

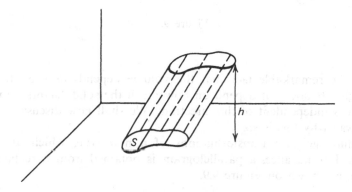

Figure 9.7

A cylinder whose base is a polygon is also called a **prism** (not necessarily a right prism).

Theorem 9-1. *The volume of a cylinder whose base has area B and whose height is h is given by the formula*

$$\boxed{V = Bh.}$$

Example. Let T be a right triangle in the plane, whose legs have lengths 4 cm and 5 cm. Find the volume of the prism with height 2 cm with T as a base, in some given direction.

The area of the base is $\frac{1}{2} \cdot 4 \cdot 5 = 10 \text{ cm}^2$. Hence by the formula, the volume of the prism is

$$V = 2 \cdot 10 = 20 \text{ cm}^3.$$

The prism has been drawn in Figure 9.8.

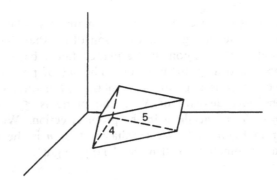

Figure 9.8

Note the remarkable fact that the volume depends *only* on the base and height. It does not depend on how much the solid "slants"; in other words, it is independent of the direction. We shall now discuss in an informal way why this is so.

We must go into transformations of a new type, which are called **shearing**. For instance, a parallelogram is obtained from a rectangle by shearing, as shown on Figure 9.9.

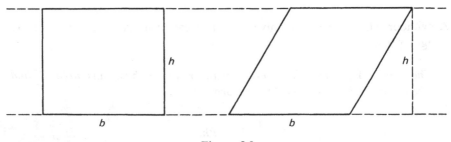

Figure 9.9

A shearing is therefore a stretching in some direction.

An informal way of thinking about the effect of shearing in 3-space is to consider a deck of cards:

Figure 9.10

If we shear the deck, we get another prism consisting of the same deck of cards.

Figure 9.11

The volume of this new "slanted" prism is the same as that of the original deck.

You already know that the area of the parallelogram with height h and side b is bh. This is the same as the area of the rectangle with sides b and h. Hence we see that for rectangles,

the area remains the same under a shearing transformation
(in the direction of one of the sides).

By approximating an arbitrary plane region by rectangles, one can then deduce the following general principle:

Theorem 9-2. *The area of a region in the plane is unchanged under shearing transformations.*

Example. Let T be a triangle with base b and height h. Any other triangle with base b and height h can be viewed as a shearing of T, as illustrated on Figure 9.12. We know that the areas of such triangles is the same, namely $\frac{1}{2}bh$.

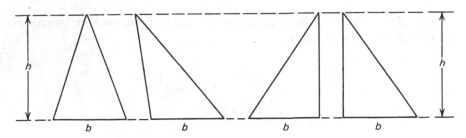

Figure 9.12

The same principle applies to volumes. We also have shearing transformations of 3-space, and the principle states:

Theorem 9-3. *The volume of a region in 3-space is unchanged under shearing transformations.*

Example. Let S be a rectangle in the plane. A prism whose base is S, and having height h is called a **parallelepiped**. All such prisms are transformed from each other by shearing, as illustrated on Figure 9.13.

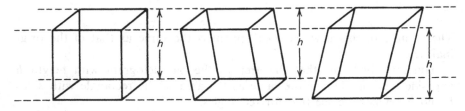

Figure 9.13

9, §1. EXERCISES

1. Lines L_1 and L_2 are parallel. Explain why all triangles ABX have the same area.

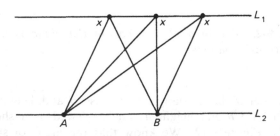

Figure 9.14

2. Find the volume of each parallelepiped below:

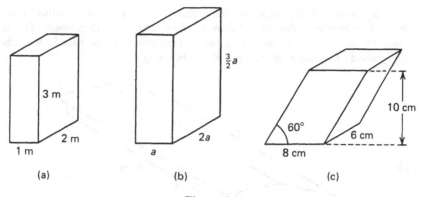

(a)   (b)   (c)

Figure 9.15

3. The **surface area** of a solid is the total of the areas of all its faces. Find the surface area of each parallelepiped above. (*Note*: You'll need to recall the dimensions of the 30–60–90 triangle for *c*.)

4. If the length of each edge of a cube is doubled, by what factor is the volume increased? What if the edge is tripled how do the volume and surface area change? What if the edge is multiplied by a factor of *r*?

5. Find the volumes of the cylinders below:

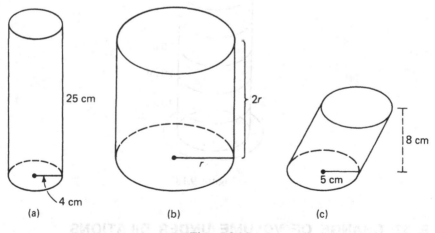

(a)   (b)   (c)

Figure 9.16

6. Suppose we have a tin can with a base of given radius, and a given height. If we double the radius and halve the height to form a new can, will the volume of the new can be larger, smaller, or the same as the original?

7. A prism has a regular hexagon of side 4 cm as base and a height of 10 cm. Find its volume and surface area.

8. Heating costs are dependent on the volume of air contained within a structure. The building illustrated below is 45 m long, and 15 m wide. On both sides the eaves are 3 m high, and the highest point of the roof is 5 m above the ground. Calculate the volume of the building.

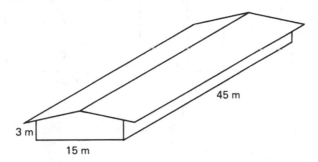

3 m

15 m

45 m

Figure 9.17

9. A graduated cylinder used for measuring volumes of liquid has a base of radius 3 cm. If the markings on the side of the cylinder are to indicate volume in units of 10 ml, how far apart should the marking lines be?

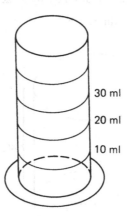

30 ml

20 ml

10 ml

Figure 9.18

9, §2. CHANGE OF VOLUME UNDER DILATIONS

How does volume change under a dilation of space by a factor of r? We know that area changes by a factor of r^2. We shall now see that volume changes by a factor of r^3. Let us first look at a rectangular box, with sides a, b, c as shown on Figure 9.19.

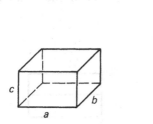
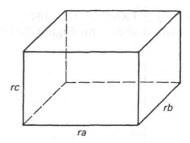

Figure 9.19

Its volume is $V = abc$. The sides of the dilated box have lengths ra, rb, rc. Hence the dilated box has volume

$$rarbrc = r^3abc = r^3V.$$

Thus the volume of the box changes by a factor of r^3.

An arbitrary region in space can be approximated by small boxes, just as we approximated area by small rectangles in Chapter 8, §2. Thus a similar statement holds for arbitrary regions in space, and we state this as a theorem.

Theorem 9-4. *Let R be a region in space, with volume V. Let D be the dilation of space by a factor of r in each one of the three perpendicular directions. Then the volume of D(R) is r^3V.*

For some applications, we have to consider dilations by different factors in different directions.

We first look at generalized dilations in the plane. Instead of dilating a region by a factor of r in all directions, we may dilate the region by this factor only in one direction. For example, if we have a rectangle with sides a, b and stretch the rectangle in the direction of one side by a factor of r, we obtain a rectangle with sides ra, b as shown on Figure 9.20.

Figure 9.20

The area of the original rectangle is ab, while the area of the partially dilated rectangle is rab.

The easiest way to assign a direction is by means of the perpendicular axes. Consider a rectangle R as shown in Figure 9.21(a). Let D be the dilation by a factor of 3 in the y-axis direction. Then the dilated rectangle $D(R)$ is shown on Figure 9.21(b).

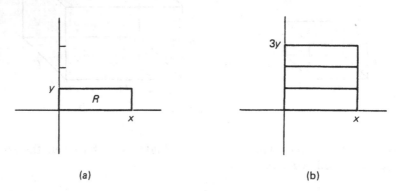

(a)　　　　　　　　　　　　　　　(b)

Figure 9.21

If A is the area of R then the area of $D(R)$ is $3A$.

By approximating an arbitrary region in the plane by means of rectangles, we are led to the following general principle.

Theorem 9-5. *Let R be a region in the plane with area A. Let D be the dilation of the plane in one of the perpendicular directions of a coordinate axis, by a factor of r. Then the area of $D(R)$ is rA.*

Let us now consider what happens to volumes in 3-space. Let R be a rectangular box with sides a, b, c as shown on Figure 9.22(a). Let us make a dilation D of space only in one direction, by a factor of r, for instance, in the direction of the first side. Then the dilated rectangular box $D(R)$ is shown on Figure 9.22(b). The volume of R is abc, and the volume of $D(R)$ is $rabc$. Thus we see that for this partial dilation only in one direction, the volume changes by a factor of r.

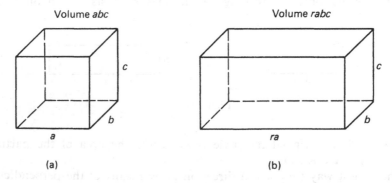

(a)　　　　　　　　　　　　　　　(b)

Figure 9.22

We can express a mixed dilation in terms of coordinates. Suppose we have a rectangular box as shown in Figure 9.23(a), and we make a dilation by a factor of 3 in the z-direction. Then we obtain the box illustrated in Figure 9.23(b).

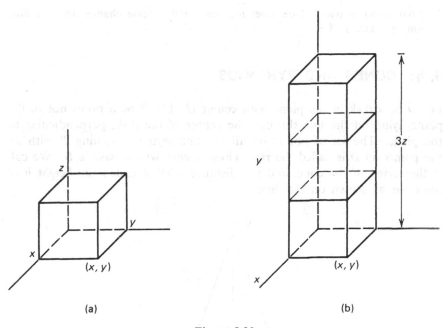

(a)   (b)

Figure 9.23

We see that the volume of the dilated box is equal to three times the volume of the original box.

In general, if we make a dilation by a factor of r in *one* of the three perpendicular directions, say in the z-direction, then the volume of a rectangular box changes by a factor of r. Indeed, the volume of the box in Figure 9.23(a) is xyz, and the volume of the dilated box is $xy(rz) = rxyz$.

By approximating an arbitrary solid region in 3-space by rectangular boxes, we find the general principle:

Theorem 9-6. *Let R be a region in space, with volume V. Let D be the dilation of space by a factor of r in one of the three perpendicular directions. Then the volume of D(R) is rV.*

In the next sections, we shall apply Theorems 9-5 and 9-6 to find the volumes contained by classical figures like cones and the sphere.

9, §2. EXERCISES

1. Suppose you have a solid S in 4-dimensional space, and S has 4-dimensional volume V. Make a dilation of 4-space by a factor of r in all directions. What is the volume of the dilated solid?

2. What about n-space? How does n-dimensional volume change under a dilation by a factor of r?

9, §3. CONES AND PYRAMIDS

Let D be a disk in the plane with center O. Let P be a point not in the plane, lying on the line through the center of the disk, perpendicular to the plane. The set of points on all the line segments joining P with all the points on D is called the **right circular cone** whose base is D. We call P the **vertex** of the cone, and the distance $|PO|$ is called the **height** h of the cone, as shown on the figure.

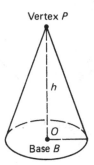

Figure 9.24

Theorem 9-7. *The volume V of the cone is given by the formula*

$$V = \tfrac{1}{3}Bh,$$

where B is the area of the base.

We shall first work out some examples where you can find the answer by using the formula. We shall then give reasons why the formula is true.

Example. Find the volume of a right circular cone whose base has radius 4 and whose height is 5.

The base has area 16π, and hence by the formula, the volume is equal to

$$\frac{1}{3}16\pi \cdot 5 = \frac{80\pi}{3}.$$

The formula for volumes just given is true for even more general figures, which are generalized cones. Let S be a region in the plane, and let P be a point not in the plane. By the **(generalized) cone** with base S and vertex P we mean the set of points which lie on the segments between P and all the points of S. We illustrate such a generalized cone in Figure 9.25.

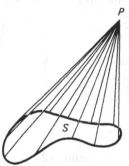

Figure 9.25

If the base is any polygon, then the (generalized) cone having this base is also called a **pyramid**. The most usual pyramids in Egypt have a square base, as shown on Figure 9.26, and are right pyramids. This means that the line through P perpendicular to the base passes through the center of the square.

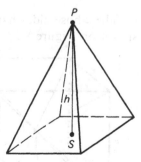

Figure 9.26

The perpendicular distance from P to the plane in which S lies is called the **height** of the cone.

Theorem 9-8. *Let a generalized cone or pyramid have base B and height h. Then the volume is*

$$V = \tfrac{1}{3}Bh.$$

Example. Find the volume of a pyramid whose base is a square with sides of length 3 m, and whose height is 5 m.

By the formula, the volume is

$$V = \tfrac{1}{3}\, 3^2 \cdot 5 = 15 \text{ m}^3.$$

This is the right answer whether we deal with a right pyramid or not. In other words, it applies to the pyramids drawn in Figure 9.27(a) and (b).

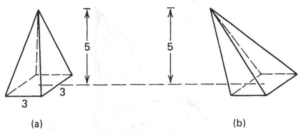

(a)   (b)

Figure 9.27

We shall now prove Theorem 9-8. The idea is first to prove a special case, and then to derive the formula in general by dilations and shearing. The main problem is to show where the $\frac{1}{3}$ comes from. Before you look at what comes next, try to work out a special case where you will discover this $\frac{1}{3}$ yourself. We shall in fact show two special cases where we can see the factor $\frac{1}{3}$ clearly. These special cases are based on situations with lots of symmetry.

Special Case. Consider a cube whose sides have length a, and let P be the center of the cube as shown on Figure 9.28.

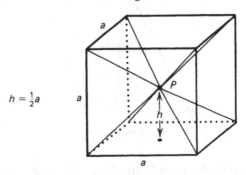

$h = \tfrac{1}{2}a$

Figure 9.28

From the center P draw the line segments to the vertices of the cube. The cube has six faces. In this way, the cube is decomposed into six pyramids, each face being a base of one of the pyramids, and P being a common vertex. By symmetry, these pyramids all have the same volume V. Since the volume of the cube is a^3, we conclude that

$$V = \tfrac{1}{6}a^3.$$

However, the height h of each pyramid is $\tfrac{1}{2}a$, and the area of the base of each pyramid is the area of each face of the cube, namely a^2. Hence we see that the volume satisfies the formula

$$V = \tfrac{1}{3}Bh = \tfrac{1}{3}a^2 \, \tfrac{1}{2}a.$$

Now apply a dilation of space by a factor of r in the direction perpendicular to the plane. Then the dilated pyramid has the same base S, and has height rh, as shown on Figure 9.29(b).

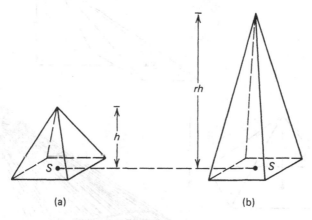

Figure 9.29

By the general principle, the volume of the dilated pyramid is equal to

$$rV = \tfrac{1}{3}Brh.$$

Hence if the formula for the volume of a pyramid with square base is known for some height, it is then true for every height by using such a dilation in the direction perpendicular to the base.

Since the formula for the volume of a right pyramid was verified in the one case when the base is a square with side a, and the height has length $a/2$ in an example of the previous section, it now follows that the formula for the volume of a right pyramid is true whenever the base is a square, and the height is arbitrary. Thus we have proved:

If a right pyramid has square base B and height h then the volume of the pyramid is

$$V = \tfrac{1}{3}Bh.$$

By shearing, we conclude that the formula for the volume of a pyramid with square base is true, whether it is a right pyramid or not. By approximating an arbitrary base by squares, one can then see that the formula must be true in general for any base, as illustrated on Figure 9.30.

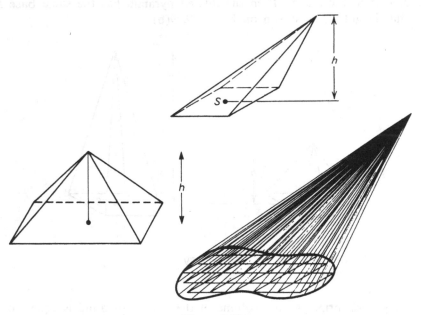

Figure 9.30

Another Special Case. There is another way of finding a special case of pyramid for which the volume formula can be verified directly. Namely, we consider a pyramid whose base is one side of a cube, and whose vertex is one of the opposite corners of the cube, as shown on Figure 9.31.

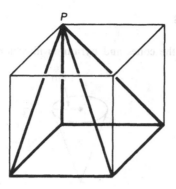

Figure 9.31

By symmetry, this pyramid has a volume which is equal to one-third the volume of the cube. Indeed, if we let P be the vertex of the pyramid, then we can construct two other pyramids, with vertex at P and whose bases are the opposite sides, namely the front side of the cube, and the right-hand side of the cube. These three pyramids cover the whole cube. Here again, if V denotes the volume of the pyramid then

$$V = \tfrac{1}{3}Bh,$$

However, the pyramid is *not* a right pyramid like an Egyptian pyramid, because its vertex does not lie just above the center of the base, but lies vertically above one of the corners of the base.

Once we have this example, we can argue exactly as before. By dilations, we get the formula for the volume for any pyramid whose vertex lies vertically above one of the corners of the base. By shearing we then get the formula for the volume of any pyramid, including a right pyramid.

Question. Do you have a preference for either one of the special cases which can be used at the beginning of the proof of Theorem 9-8? We do not. Discuss your preferences in class, and give reasons for them. For instance, with the first special case, we could get the volume of a right pyramid using only dilations, without using shearing. With the second special case, we must use shearing to get the volume of a right pyramid. On the other hand, it may seem clearer in the second special case where the factor $\tfrac{1}{3}$ comes from. Also the second special case will be used in the proof of Theorem 9-9 below.

9, §3. EXERCISES

1. Find the volumes of the cones and pyramids illustrated:

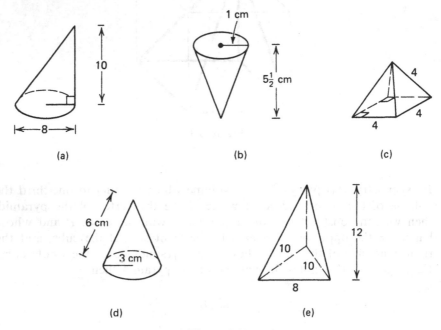

Figure 9.32

2. If the radius of the base of a cone is doubled, by what factor is the volume of the cone increased?

3. The Egyptian pyramids were built as tombs for the sovereigns of the Old and Middle Kingdoms. The largest, that of Khufu, is $220 \, m^2$ at the base, and 146.59 m high. Calculate its volume. How many blocks in the shape of cubes 1 meter on a side would be necessary to build such a monument (approximately)?

4. Find the total surface area and volume of a pyramid whose base is a regular hexagon base with edge 8; and whose height is 10.

5. Derive the formula for the volume of a pyramid whose sides are equilateral triangles with side s, and whose base is a square.

6. Suppose we have a pyramid with triangular base. Consider a cross section of the pyramid obtained by slicing it with a plane parallel to the base. Show that this triangular cross section is similar to the base. [*Hint*: First apply the similarity theorems to each face.]

7. Consider a pyramid with triangular base and height H. Suppose we obtain a cross section as shown in the figure by slicing the pyramid at a distance h

from the top. Show that the ratio of the areas of the cross section triangle to the base is h^2/H^2.

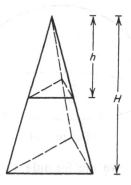

Figure 9.33

8. A pyramid has an equilateral triangle of side 6 as a base, and has height 12. Find the volume of the pyramid obtained by cutting off the top of the original pyramid at a distance of 4 units from the top. Find the volume of the remaining solid with trapezoidal sides (called a **frustrum**).

9, §4. THE VOLUME OF THE BALL

Let P be a given point in space, and let r be a number > 0. We define the **sphere** of radius r and center P to be the set of all points in space whose distance from P is equal to r. We define the **ball** of radius r and center P to be the set of all points in space whose distance from P is less than or equal to r.

We shall next determine the volume of the ball.

Theorem 9-9. *The volume of a ball of radius r is $\frac{4}{3}\pi r^3$.*

Proof. Suppose we know the volume V of a ball of radius 1. Then by Theorem 9-7 the volume of the ball of radius r is equal to

$$Vr^3.$$

So it suffices to prove that the volume of the ball of radius 1 is $4\pi/3$.

As before, our method is to approximate the ball by simpler regions whose volume we already know. Such regions are cylinders, and we know that

Volume of cylinder of radius r and height $h = \pi r^2 h$.

Instead of the ball, it suffices to prove the formula for the volume of the upper half of the ball of radius 1, as shown on the figure.

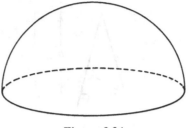

Figure 9.34

Thus it will suffice to prove the formula for the half ball of radius 1, namely

Volume of the half ball = $\frac{2}{3}\pi$.

We cut the half ball into slices, like this:

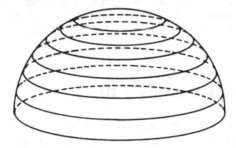

Figure 9.35

We approximate these slices with cylinders. In the next figure, we show both a cross section of the cylinders, and a perspective figure of the cylinders themselves.

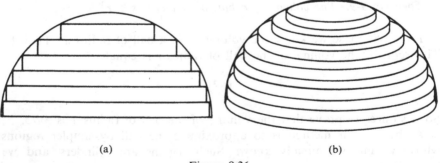

(a)   (b)

Figure 9.36

We need to determine the radii and the heights of the cylinders. Then we can write down their volumes, add these up, and the sum of these volumes approximates the volume of the half ball. Let us first draw a special case of six cylinders.

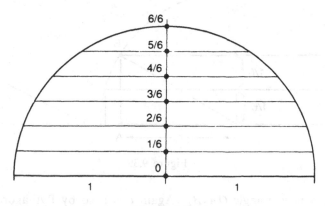

Figure 9.37

We make all the cylinders of the same height. Since the ball has radius 1, each height is equal to $\frac{1}{6}$. If instead of six cylinders we use an arbitrary number n, all of the same height, then each cylinder has height $1/n$.

Now for the radius of each cylinder, let us again look first at the case of six cylinders. We shall use the Pythagoras theorem. Let us draw the first cylinder at the bottom.

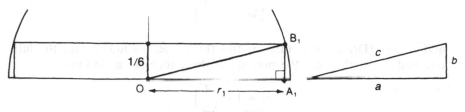

Figure 9.38

We have a right triangle OA_1B_1 in Figure 9.38, and in a right triangle we know that

$$a^2 + b^2 = c^2.$$

Let r_1 be the radius of the bottom cylinder. Here $c = 1$, so

$$r_1^2 + \left(\frac{1}{6}\right)^2 = 1^2 = 1$$

because 1 is the radius of the ball. Hence

$$r_1^2 = 1 - \left(\frac{1}{6}\right)^2.$$

Next we look at the second cylinder on Figure 9.39.

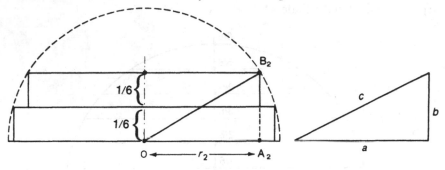

Figure 9.39

We have a new triangle OA_2B_2. Again $c = 1$, so by Pythagoras

$$r_2^2 + \left(\frac{2}{6}\right)^2 = 1$$

and

$$r_2^2 = 1 - \left(\frac{2}{6}\right)^2.$$

We continue in the same way. For the third cylinder we shall get

$$r_3^2 = 1 - \left(\frac{3}{6}\right)^2$$

and so on. (Draw the picture for this third one, the fourth one, the fifth one and the sixth one.) We may write these together as follows:

$$r_1^2 = 1 - \left(\frac{1}{6}\right)^2,$$

$$r_2^2 = 1 - \left(\frac{2}{6}\right)^2,$$

$$r_3^2 = 1 - \left(\frac{3}{6}\right)^2,$$

$$r_4^2 = 1 - \left(\frac{4}{6}\right)^2,$$

$$r_5^2 = 1 - \left(\frac{5}{6}\right)^2,$$

$$r_6^2 = 1 - \left(\frac{6}{6}\right)^2.$$

The height of each cylinder is $\frac{1}{6}$. Therefore:

The sum of the volumes of the six cylinders =

$$\pi\left(1 - \left(\frac{1}{6}\right)^2\right)\cdot\frac{1}{6} + \pi\left(1 - \left(\frac{2}{6}\right)^2\right)\cdot\frac{1}{6} + \cdots + \pi\left(1 - \left(\frac{6}{6}\right)^2\right)\cdot\frac{1}{6}.$$

Instead of six cylinders, we could take seven, eight, or an arbitrary number n. In general, suppose we have n cylinders, each one of height $1/n$. Then the radii of these cylinders are

$$r_1 = 1 - \left(\frac{1}{n}\right)^2, \quad r_2 = 1 - \left(\frac{2}{n}\right)^2, \quad \cdots \quad, \quad r_n = 1 - \left(\frac{n}{n}\right)^2.$$

Therefore

The sum of the volumes of the n cylinders =

$$\pi\left(1 - \left(\frac{1}{n}\right)^2\right)\cdot\frac{1}{n} + \pi\left(1 - \left(\frac{2}{n}\right)^2\right)\cdot\frac{1}{n} + \cdots + \pi\left(1 - \left(\frac{n}{n}\right)^2\right)\cdot\frac{1}{n}.$$

When n becomes larger and larger, the sum of the volumes of the cylinders approaches the volume of the half ball of radius 1.

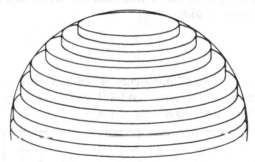

Figure 9.40

Observe that in the formula for the sum of the volumes of the n cylinders, there is a common factor π/n. We factor this out, and we obtain:

The sum of the volumes of the n cylinders

$$= \frac{\pi}{n}\left[1 - \left(\frac{1}{n}\right)^2 + 1 - \left(\frac{2}{n}\right)^2 + 1 - \left(\frac{3}{n}\right)^2 + \cdots + 1 - \left(\frac{n}{n}\right)^2\right]$$

$$= \frac{\pi}{n}\left[n - \left(\frac{1}{n}\right)^2 - \left(\frac{2}{n}\right)^2 - \cdots - \left(\frac{n}{n}\right)^2\right]$$

$$= \pi - \pi\left[\left(\frac{1}{n}\right)^2\cdot\frac{1}{n} + \left(\frac{2}{n}\right)^2\cdot\frac{1}{n} + \cdots + \left(\frac{n}{n}\right)^2\cdot\frac{1}{n}\right].$$

Now we must evaluate what this expression approaches when n becomes large. We do this in the following lemma.

Lemma. *The sum*

$$\left(\frac{1}{n}\right)^2 \cdot \frac{1}{n} + \left(\frac{2}{n}\right)^2 \cdot \frac{1}{n} + \cdots + \left(\frac{n}{n}\right)^2 \cdot \frac{1}{n} \quad approaches \quad \frac{1}{3}$$

when n becomes large.

Once we prove the lemma, we conclude that the volume of the half ball of radius 1 is

$$\pi - \frac{1}{3}\pi = \frac{2\pi}{3},$$

which is what we wanted to prove.

Now we want to prove the lemma. At first sight, this looks rather tough. There are some squares, lots of n, and it is not at all clear how the sum behaves. So what do we do?

We first consider the simpler problem without the squares, to see if we can get a better idea of what goes on. We shall return to the formula with the squares afterwards. So we write down a similar sum without the squares as follows:

$$\frac{1}{n} \cdot \frac{1}{n} + \frac{2}{n} \cdot \frac{1}{n} + \cdots + \frac{n}{n} \cdot \frac{1}{n}.$$

Can we see what this sum approaches when n becomes arbitrarily large? To see it, we consider a square of sides 1 and we cut up this square into little squares whose sides have length $1/n$.

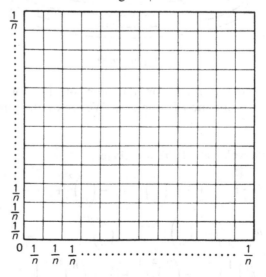

Figure 9.41

We observe that

$\dfrac{1}{n} \cdot \dfrac{1}{n}$ is the area of a square whose sides have length $\dfrac{1}{n}$,

$\dfrac{2}{n} \cdot \dfrac{1}{n}$ is the area of a rectangle whose sides have lengths $\dfrac{2}{n}$ and $\dfrac{1}{n}$,

$\dfrac{3}{n} \cdot \dfrac{1}{n}$ is the area of a rectangle whose sides have lengths $\dfrac{3}{n}$ and $\dfrac{1}{n}$,

. . . .

$\dfrac{n}{n} \cdot \dfrac{1}{n}$ is the area of a rectangle whose sides have lengths $\dfrac{n}{n}$ and $\dfrac{1}{n}$.

We can draw these rectangles as on a staircase.

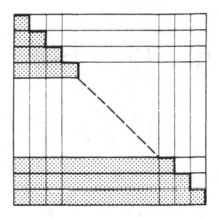

Figure 9.42

The sum without the squares is just the area of the staircase. If we increase n, if n gets bigger and bigger, then the staircase has more steps, and its area approaches half the square. In other words, we have solved our problem for the simpler sums, namely:

As n becomes larger and larger, the sum

$$\frac{1}{n} \cdot \frac{1}{n} + \frac{2}{n} \cdot \frac{1}{n} + \cdots + \frac{n}{n} \cdot \frac{1}{n} \textbf{ approaches } \frac{1}{2}.$$

This was not so bad, so we are now ready to try the original sum which we encountered in the lemma. Instead of a square, we draw a cube with sides 1, and we slice it up into little cubes of sides $1/n$.

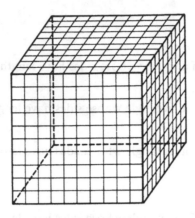

Figure 9.43

In the sum

$$\left(\frac{1}{n}\right)^2 \cdot \frac{1}{n} + \left(\frac{2}{n}\right)^2 \cdot \frac{1}{n} + \cdots + \left(\frac{n}{n}\right)^2 \cdot \frac{1}{n}$$

the first term is

$$\left(\frac{1}{n}\right)^2 \cdot \frac{1}{n} = \frac{1}{n^3}.$$

This is the volume of a cube of sides $1/n$.

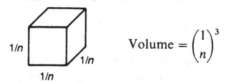

$$\text{Volume} = \left(\frac{1}{n}\right)^3$$

Figure 9.44

The next term in the sum is

$$\left(\frac{2}{n}\right)^2 \cdot \frac{1}{n} = \text{volume of rectangular box of height } 1/n \text{ and square base with sides } 2/n.$$

We draw this box just below the first one.

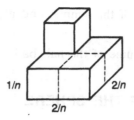

Figure 9.45

The third term is

$$\left(\frac{3}{n}\right)^2 \cdot \frac{1}{n} = \text{volume of rectangular box of height } 1/n \text{ and}$$
square base of sides $3/n$.

And so on, we go like this up to n/n.

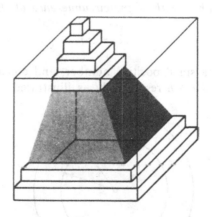

Figure 9.46

So the picture illustrating the sum in the lemma is that of a 3-dimensional staircase.

When n becomes bigger and bigger, the stairs approach a pyramid, with a square base and height 1. The volume of this pyramid is one-third of the volume of the cube. Since the cube has volume 1, the volume of this pyramid is 1/3. Therefore we find:

As n becomes bigger and bigger, the sum

$$\left(\frac{1}{n}\right)^2 \cdot \frac{1}{n} + \left(\frac{2}{n}\right)^2 \cdot \frac{1}{n} + \cdots + \left(\frac{n}{n}\right)^2 \cdot \frac{1}{n} \quad \textbf{approaches} \quad \frac{1}{3}.$$

This concludes the proof of the lemma, and also concludes the proof for the formula

$$\text{Volume of the half ball} = \tfrac{2}{3}\pi.$$

9, §5. THE AREA OF THE SPHERE

To get the formula for the area of the sphere, we shall use a method which could already have been used to prove the formula for the length of a circle. This method is different from the method used in Chapter 8.

So we start with this new method to give a new proof of the formula for the length of a circle. We only assume known the formula for the area of a disc of radius r, namely

$$A = \pi r^2,$$

and we prove:

Theorem 9-10. *The length of the circumference of the circle is given by*

$$c = 2\pi r.$$

Proof. Let h be a small positive number, and consider two concentric circles of radii r and $r + h$ respectively as illustrated.

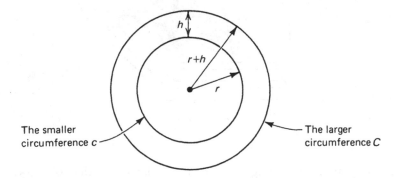

Figure 9.47

The circle of radius r is surrounded by a thin band of width h. The area of this band is the difference between the areas of the disk of radius $r + h$ and the disk of radius r. Let us denote by B the area of the band of width h. Then we can write the equation

$$B = \pi(r + h)^2 - \pi r^2$$
$$= \pi(r^2 + 2rh + h^2) - \pi r^2$$
$$= 2\pi rh + \pi h^2.$$

If instead of a circular band we considered a straight rectangle of length *L* and width *h* then the area of the band would be *Lh.*

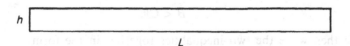

Figure 9.48

Because our band is circular, instead of an equality, we have two in-equalities for its area. First, consider a rectangle of width *h* and length *c*, where *c* is the length of the small circle. We wrap this rectangle around the circle as shown on Figure 9.49.

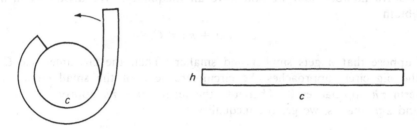

Figure 9.49

If we curve the rectangle and wrap it around the circle, then the inner side of the rectangle sticks exactly around the smaller circle, but we have to stretch it so that the outer side fits around the bigger circle. This means that the area of the rectangle is smaller than the area of the band between the two circles. We can write this as an inequality:

$$hc < B.$$

Similarly, let *C* be the circumference of the big circle. We now take a rectangle whose length is *C* and whose width is *h*, and we wrap this rect-angle inside the larger circle.

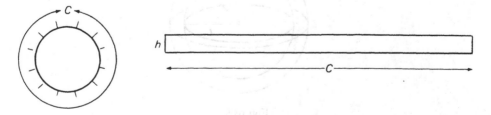

Figure 9.50

Since we have to shrink the inner edge of the rectangle, the area of the rectangle is bigger than the area of the band, and we can write the inequality

$$B < Ch.$$

We then write the two inequalities together in the form

$$ch < B < Ch.$$

Substituting the exact expression for B that we found previously, we have

$$ch < 2\pi rh + \pi h^2 < Ch.$$

You should know that we can divide both sides of an inequality by a positive number, and we still have an inequality. We divide by h, and obtain

$$c < 2\pi r + \pi h < C.$$

Suppose that h gets smaller and smaller. Then the circumference C of the big circle approaches the circumference c of the small circle. The term πh approaches 0. Therefore the term $2\pi r + \pi h$ approaches both c and $2\pi r$, that is, we get the inequality

$$c \leqq 2\pi r \leqq c.$$

This proves that $c = 2\pi r$, and concludes the proof of Theorem 9-10.

Now we go back to the sphere, and we shall prove:

Theorem 9-11. *Let A be the area of the sphere. Then*

$$A = 4\pi r^2.$$

Proof. Again we let h be a small positive number, but now we consider two concentric spheres of radii r and $r + h$ respectively, as illustrated.

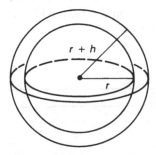

Figure 9.51

Let s be the area of the sphere of radius r, and let S be the area of the sphere of radius $r + h$. Let v be the volume of the ball of radius r, and let V be the volume of the larger ball of radius $r + h$. The volume of the thin shell between the two balls is then equal to the difference:

$$\text{Volume of shell} = V - v = \tfrac{4}{3}\pi(r + h)^3 - \tfrac{4}{3}\pi r^3,$$

using the formula for the volume of the ball proved in the previous section. You should know from algebra (see below) that

$$(r + h)^3 = r^3 + 3r^2h + 3rh^2 + h^3.$$

Hence the above difference is equal to:

$$\begin{aligned}
\text{Volume of shell} &= \tfrac{4}{3}\pi[(r + h)^3 - r^3] \\
&= \tfrac{4}{3}\pi[r^3 + 3r^2h + 3rh^2 + h^3 - r^3] \\
&= \tfrac{4}{3}\pi[3r^2h + 3rh^2 + h^3] \\
&= \tfrac{4}{3}\pi[3r^2 + 3rh + h^2]h
\end{aligned}$$

after we factor out h in this last step.

On the other hand, the volume of the shell is approximately equal to the area of the sphere times the width of the shell, and this width is h. More precisely, we have the same inequality as for the area of the band around the circle. The volume of the shell is larger than the area of the small sphere times h; and it is smaller than the area of the big sphere times h. We can write these inequalities in the form:

$$hs < \text{volume of shell} < hS.$$

Substituting the expression we found for the volume of the shell, we can write

$$hs < \tfrac{4}{3}\pi[3r^2 + 3rh + h^2]h < hS.$$

We divide by h to find

$$s < \tfrac{4}{3}\pi[3r^2 + 3rh + h^2] < S.$$

As h approaches 0, the area of the big sphere S approaches the area of the small sphere s. The expression in the middle approaches

$$\tfrac{4}{3}\pi 3r^2,$$

which is precisely $4\pi r^2$. This concludes the proof that the area of the sphere of radius r is $4\pi r^2$.

Remark. The formula

$$(r + h)^3 = r^3 + 3r^2h + 3rh^2 + h^3$$

is quite simple to prove. First, we have

$$(r + h)^2 = (r + h)(r + h) = r(r + h) + h(r + h)$$
$$= r^2 + rh + hr + h^2$$
$$= r^2 + 2rh + h^2.$$

Then

$$(r + h)^3 = (r + h)(r + h)^2 = r(r^2 + 2rh + h^2) + h(r^2 + 2rh + h^2)$$
$$= r^3 + 2r^2h + rh^2 + hr^2 + 2rh^2 + h^3$$
$$= r^3 + 3r^2h + 3rh^2 + h^3,$$

as desired.

9, §5. EXERCISES

1. Write out and prove the expansion for $(a + b)^4$, using the formula for $(a + b)^3$.
2. Write out and prove the expansion for $(a + b)^5$, using the formula for $(a + b)^4$.

CHAPTER 10

Vectors and Dot Product

Consider an airplane flying through a crosswind.

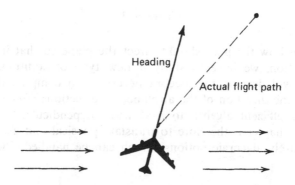

Figure 10.1

While the pilot keeps the plane facing in a certain direction, the action of the wind alters the actual course of the plane. In this chapter we shall use some geometry to describe mathematically objects which represent the force of the wind and the direction of the airplane. Just as in the picture, we represent these by arrows, as on the figure.

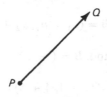

Figure 10.2

The distinctive thing about the arrow is its beginning point P and its end point Q. Thus we define a **located vector** to be a pair of points $\overrightarrow{PQ}$, where P is the beginning point and Q is the end point. The length of the arrow, that is the length of the segment $|PQ|$, represents the magnitude, and the arrow points in the given direction.

We can represent the force of the wind using a vector; the direction of the vector is the direction of the wind, the length of the vector represents the amount of force (speed) of the wind. For example, we can represent a westerly wind of 50 km/hr by the vector $\overrightarrow{OW}$, where $W = (50, 0)$, as shown on the figure.

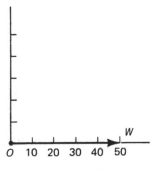

Figure 10.3

To determine how the pilot should direct the plane so that it flies in the desired direction, we have to study a new type of addition. Thus the first section will develop the algebra necessary to compute the effect of the wind on the direction of the airplane. The sections after that will develop a very efficient algebra to deal with perpendicularity. The main point of this chapter is therefore to translate physical and geometric concepts into purely algebraic notions, which can be handled efficiently.

10, §1. VECTOR ADDITION

Let A and B be points in the plane. We write their coordinates,

$$A = (a_1, a_2) \quad \text{and} \quad B = (b_1, b_2).$$

We define their **sum** $A + B$ to be

$$A + B = (a_1 + b_1, a_2 + b_2).$$

Example. Let $A = (1, 4)$ and $B = (-1, 5)$. Then

$$A + B = (1 - 1, 4 + 5) = (0, 9).$$

Example. Let $A = (-3, 6)$ and $B = (-2, -7)$. Then

$$A + B = (-3 - 2, 6 - 7) = (-5, -1).$$

This addition satisfies properties similar to the addition of numbers—and no wonder, since the coordinates of a point are numbers. Thus we have for any points A, B, C:

Commutativity. $A + B = B + A$.

Associativity. $A + (B + C) = (A + B) + C$.

Zero element. Let $O = (0, 0)$. Then $A + O = O + A = A$.

Additive inverse. If $A = (a_1, a_2)$, then the point

$$-A = (-a_1, -a_2)$$

is such that

$$A + (-A) = O.$$

These properties are immediately proved from the definitions. For instance, let us prove the first one. We have:

$$A + B = (a_1 + b_1, a_2 + b_2) = (b_1 + a_1, b_2 + a_2) = B + A.$$

Our proof simply reduces the property concerning points to the analogous property concerning numbers. The same principle applies to the other properties. Note especially the additive inverse. For instance,

$$\text{if } A = (2, -5), \quad \text{then} \quad -A = (-2, 5).$$

As with numbers, we shall write $A - B$ instead of $A + (-B)$.

Example. If $A = (2, -5)$ and $B = (3, -4)$, then

$$A - B = (2 - 3, -5 - (-4)) = (-1, -1).$$

We shall now interpret this addition and subtraction geometrically. We consider examples.

Example. Let $A = (1, 2)$ and $B = (3, 1)$. To find $A + B$, we start at A, go 3 units to the right, and 1 unit up, as shown in Figure 10.4(a) and (b).

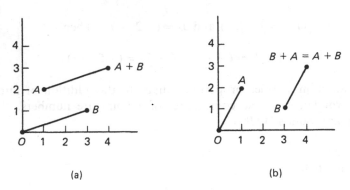

(a)   (b)

Figure 10.4

Thus we see geometrically that $A + B$ is obtained from A by the same procedure as B is obtained from O. In this geometric representation, we also see that the line segment between A and $A + B$ is parallel to the line segment between O and B, as shown in Figure 10.4(a). Similarly, the line segment between O and A is parallel to the line segment between B and $A + B$. Thus the four points

$$O, \quad A, \quad B, \quad A + B$$

form the four corners of a parallelogram, which we draw in Figure 10.5.

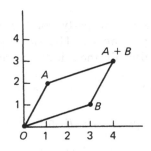

Figure 10.5

This gives a geometric interpretation of addition, and is sometimes referred to as the **parallelogram law** for addition.

Next, we consider subtraction. Let $A - B = C$. Then $B + C = A$. Thus $A - B$ is the point C at one corner of the parallelogram illustrated in Figure 10.6 such that $B + C = A$.

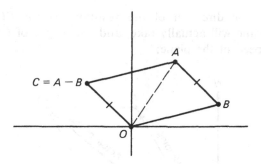

Figure 10.6

We find this point $C = A - B$ by drawing a segment $\overrightarrow{OC}$, starting at the origin O, parallel to $\overrightarrow{BA}$, and having the same length, in the same direction as $\overrightarrow{BA}$. Pick several numerical values for A and B, and draw A, B, $A - B$ to check this in special cases.

Example. We may now return to the example given at the beginning.

Suppose a pilot sets a course of 30° East of North at a speed of 500 km/hr. What will his actual course be, considering the effect of the westerly wind blowing at 50 km/hr? We first represent the plane's direction and speed using another vector $\overrightarrow{OP}$, where the length of $\overrightarrow{OP}$ is 500, and the direction of $\overrightarrow{OP}$ is 30° off the vertical (North):

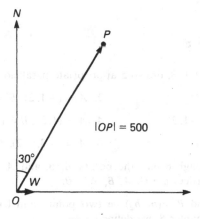

Figure 10.7

It is a law of physics that the resulting force acting on the airplane is the "vector sum" of the vectors representing the wind and the heading of the plane. This sum is found by drawing a parallelogram and locating point $P + W$. The direction of the resultant vector $\overrightarrow{O(P + W)}$ is the direction the plane will actually take, and the length of $\overrightarrow{O(P + W)}$ represents the net speed of the plane:

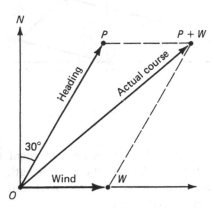

Figure 10.8

A navigator can accurately plot the vectors P and W, and find the resultant $P + W$ using a set of drafting instruments. He would then measure the length of $\overrightarrow{O(P + W)}$ and the angle it makes with true North to determine the actual path and speed of the plane. This is the procedure used when navigating "by hand" aboard a plane or boat.

10, §1. EXERCISES

Plot the points A, B, $A + B$, drawing appropriate parallelograms.

1. $A = (1, 4)$, $B = (3, 2)$ 2. $A = (-1, 2)$, $B = (3, 1)$

3. $A = (-1, 1)$, $B = (-1, 2)$ 4. $A = (1, 5)$, $B = (1, 1)$

5. $A = (-2, 1)$, $B = (1, 2)$ 6. $A = (-3, -2)$, $B = (-1, -1)$

7. In Exercises 1 through 6 plot the points A, B, and $A - B$. Draw the parallelogram whose vertices are O, A, B, $A - B$.

8. Let $A = (a_1, a_2)$ and $B = (b_1, b_2)$ be two points, and let c be a number. Remember that in Chapter 8, we defined cA.
   (a) What are the coordinates of the point cA? of point cB? of $cA - cB$?
   (b) What are the coordinates of the point $A - B$?
   (c) Show that the point $cA - cB$ is equal to the point $c(A - B)$.

9. Let $A = (1, 2)$ and $B = (3, 1)$. On a sheet of graph paper, draw $A + B$, $A + 2B$, $A + 3B$, $A - 2B$, $A - 3B$, $A + 4B$, $A - 4B$. What do you observe about the position of all these points?

10. Let $A = (2, -1)$ and $B = (-1, 1)$. Draw the points $A + B$, $A + 2B$, $A + 3B$, $A + 4B$, $A - B$, $A - 2B$, $A - 3B$, $A - 4B$, $A + \frac{1}{2}B$, $A - \frac{1}{2}B$. Again, what do you observe about the position of all these points?

11. **Exercise in 3-space.**
    (a) Define a point in 3-dimensional space to be a triple of numbers, for instance, $A = (3, 1, -2)$. If $B = (-1, 4, 5)$, how would you add $A + B$ in analogy with the 2-dimensional case?
    (b) In general, if $A = (a_1, a_2, a_3)$ and $B = (b_1, b_2, b_3)$, how would you add $A + B$?

10, §2. THE SCALAR PRODUCT

In §1 we introduced the idea of adding and subtracting points using coordinates. Many students ask whether there is some way we can "multiply" points that makes sense. The answer is yes, and we develop the theory by first defining a product algebraically, and then later by examining what its geometric implications are.

Let $A = (a_1, a_2)$ and let $B = (b_1, b_2)$. We define

$$A \cdot B = a_1 b_1 + a_2 b_2.$$

Thus $A \cdot B$ is a **number**, called the **scalar**, or **dot product** of A and B.

Example. Let $A = (3, 1)$ and $B = (-5, 4)$. Then

$$A \cdot B = 3(-5) + 1 \cdot 4 = -11.$$

Thus $A \cdot B$ is the number -11.

For the moment we do not give a geometric interpretation for this product, but we shall soon see that it is a very useful concept. It satisfies properties very similar to the properties of products of numbers. We list these. They include commutativity and distributivity.

Theorem 10-1. *The scalar product satisfies the following properties:*

SP 1. *It is commutative, that is* $A \cdot B = B \cdot A$.

SP 2. *If* A, B, C *are points, then distributivity holds, namely*

$$A \cdot (B + C) = A \cdot B + A \cdot C = (B + C) \cdot A.$$

SP 3. *If* x *is a number, then*

$$(xA) \cdot B = x(A \cdot B) = A \cdot (xB).$$

SP 4. *If $A = O$, then $A \cdot A = 0$, and if $A \neq O$, then $A \cdot A > 0$.*

Proof. Concerning the first, we have

$$A \cdot B = a_1 b_1 + a_2 b_2 = b_1 a_1 + b_2 a_2 = B \cdot A.$$

Note how our proof reduces **SP 1** to the commutativity property for numbers.

For **SP 2**, let $C = (c_1, c_2)$. Then

$$B + C = (b_1 + c_1, b_2 + c_2)$$

and

$$A \cdot (B + C) = a_1(b_1 + c_1) + a_2(b_2 + c_2)$$
$$= a_1 b_1 + a_1 c_1 + a_2 b_2 + a_2 c_2.$$

Rearranging the terms yields

$$a_1 b_1 + a_2 b_2 + a_1 c_1 + a_2 c_2,$$

which is none other than $A \cdot B + A \cdot C$. This proves **SP 2**.

We leave the proof of **SP 3** as an exercise.

Finally, for **SP 4** we observe that if $A \neq O$, then one of the coordinates a_1 or a_2 is not 0, and its square is > 0. Hence

$$a_1^2 + a_2^2 > 0$$

because one of a_1^2 or a_2^2 is > 0. This proves **SP 4**.

From the commutativity and distributivity, we can derive formulas as we did with numbers. For instance, let us use the notation

$$A \cdot A = A^2.$$

Then with this notation, we have the usual formulas:

$$(A + B)^2 = A^2 + 2A \cdot B + B^2, \qquad (A - B)^2 = A^2 - 2A \cdot B + B^2,$$

$$(A + B) \cdot (A - B) = A^2 - B^2.$$

Let us prove one of these formulas, say the second one, and leave the others as exercises. We have:

$$(A - B) \cdot (A - B) = A \cdot (A - B) - B \cdot (A - B)$$
$$= A \cdot A - A \cdot B - B \cdot A + B \cdot B$$
$$= A \cdot A - 2A \cdot B + B \cdot B$$
$$= A^2 - 2A \cdot B + B^2.$$

If you remember the proof of the analogous formula for numbers, you will realize that it is the same proof that we have given above. Also note that we have not used coordinates explicitly. We have only used properties **SP 1** through **SP 4**.

Warning. *We do not write A^3, or any other power of A, except A^2 as above.* Also, we do not take products of more than two points, i.e. we do not form $A \cdot B \cdot C$. This would not make sense.

The "square" A^2 has an interpretation in terms which we already know. We observe that

$$A^2 = a_1^2 + a_2^2$$

is exactly the square of the distance between the origin O and the point A, by the Pythagoras theorem. We shall use the symbol

$$|A|$$

to denote this distance. Thus we can write:

$$|A| = d(O, A) = \sqrt{A \cdot A} = \sqrt{a_1^2 + a_2^2}.$$

We shall call $|A|$ the **norm** of A.

Example. Find the norm of A if $A = (-2, 4)$.
By definition, the norm of A is given by

$$|A| = \sqrt{(-2)^2 + 4^2} = \sqrt{4 + 16} = \sqrt{20}.$$

Also observe that the formula for the distance between two points A, B can be written in the form

$$d(A, B) = |A - B| = \sqrt{(A - B) \cdot (A - B)}.$$

10, §2. EXERCISES

1. Find the scalar product of the following pairs A, B.
   (a) $A = (1, 3)$ and $B = (-1, 5)$
   (b) $A = (-4, -2)$ and $B = (-3, 6)$
   (c) $A = (3, 7)$ and $B = (2, -5)$
   (d) $A = (-5, 2)$ and $B = (-4, 3)$
   (e) $A = (-6, 3)$ and $B = (-5, -4)$

2. Define the scalar product in the 3-dimensional case. Prove that it has the same properties as in the 2-dimensional case. Find the scalar product of the following pairs A, B.
   (a) $A = (3, 1, -1)$ and $B = (-2, 3, 4)$
   (b) $A = (-4, 1, 1)$ and $B = (2, 1, 1)$
   (c) $A = (3, -1, 5)$ and $B = (-2, 4, 2)$
   (d) $A = (4, 1, -2)$ and $B = (3, 2, 7)$

3. Using distributivity and commutativity, write out the proofs for the formulas concerning $(A + B)^2$ and $(A + B) \cdot (A - B)$.

4. Find the norm of A in each of the following cases.
   (a) $A = (5, 3)$ (b) $A = (5, -3)$ (c) $A = (-5, 3)$
   (d) $A = (-5, -3)$ (e) $A = (2, 4)$ (f) $A = (-2, 4)$
   (g) $A = (2, -4)$ (h) $A = (-2, -4)$

5. For each point A below, find a point B such that $\overline{OB}$ is perpendicular to $\overline{OA}$.
   (a) $A = (3, 1)$ (b) $A = (6, 2)$ (c) $A = (-4, -2)$
   (d) $A = (1, 0)$ (e) $A = (3, 0)$ (f) $A = (0, 1)$

6. For each point A in Exercise 5 above, find the point $P = 2A$. Is $\overline{OB}$ perpendicular to $\overline{OP}$ in each case?

7. Look over your answers to Exercises 5 and 6. Can you find a simple formula which, given points A and B, can tell if $\overline{OA}$ is perpendicular to $\overline{OB}$?

10, §3. PERPENDICULARITY

In the previous section we defined the **scalar** or **dot product**. Note that the dot product of two points $A \cdot B$ may very well be 0 without either A or B being 0.

   Example. Let $A = (3, 5)$ and $B = (10, -6)$. Then

$$A \cdot B = 30 - 30 = 0.$$

We want to find a simple geometric interpretation for the property that $A \cdot B = 0$. Ultimately, we wish to convince ourselves that this condition is equivalent to the fact that $\overline{OA}$ and $\overline{OB}$ are perpendicular. Recall

Theorem 5-1 of Chapter 5, §1 concerning the perpendicular bisector. We see that $\overrightarrow{OA}$ is perpendicular to $\overrightarrow{OB}$ if and only if

$$d(A, B) = d(A, -B)$$

as shown on Figure 10.9.

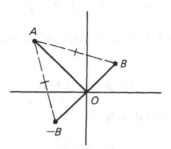

Figure 10.9

But $d(A, B) = |A - B|$, and

$$d(A, -B) = |A - (-B)| = |A + B|.$$

Thus the condition

$$|A + B| = |A - B|$$

is the one expressing perpendicularity corresponding to the corollary of the Pythagoras theorem. The next theorem will prove that this condition is equivalent to $A \cdot B = 0$. It then allows us to define A to be **perpendicular to** B if $A \cdot B = 0$, and similarly we define two vectors $\overrightarrow{OA}$ and $\overrightarrow{OB}$ to be **perpendicular** if $A \cdot B = 0$.

We write

$$\overrightarrow{OA} \perp \overrightarrow{OB} \qquad \text{or} \qquad A \perp B$$

to denote that $\overrightarrow{OA}$ is perpendicular to $\overrightarrow{OB}$, or A is perpendicular to B.

Theorem 10-2. *We have $A \cdot B = 0$ if and only if*

$$|A + B| = |A - B|.$$

Proof. Taking the square of each side, we see that

$$|A + B| = |A - B| \quad \text{if and only if} \quad |A + B|^2 = |A - B|^2.$$

We expand the right-hand side according to our definition of the dot product. We see that this condition is true if and only if

$$(A + B) \cdot (A + B) = (A - B) \cdot (A - B),$$

which is true if and only if

$$A \cdot A + 2A \cdot B + B \cdot B = A \cdot A - 2A \cdot B + B \cdot B.$$

We can cancel $A \cdot A$ and $B \cdot B$ from both sides, and we see that our condition is true if and only if

$$2A \cdot B = -2A \cdot B,$$

which, in turn, is true if and only if

$$4A \cdot B = 0.$$

This is true if and only if $A \cdot B = 0$, and our theorem is proved.

Example. Given $A = (3, 1)$, find a vector $\overrightarrow{OB}$ such that $\overrightarrow{OA}$ is perpendicular to $\overrightarrow{OB}$ (or A is perpendicular to B).

We do this by inspection. Let $B = (-1, 3)$. Then

$$A \cdot B = 3(-1) + 1(3) = 0.$$

So B is a possible solution. Observe that $(-2, 6)$, or $(-3, 9)$ are also solutions. Can you describe still other solutions? What is the general principle which leads you to find these other solutions?

We define two vectors $\overrightarrow{PQ}$ and $\overrightarrow{MN}$ to be **perpendicular** if and only if $Q - P$ is perpendicular to $N - M$, that is,

$$(Q - P) \cdot (M - N) = 0.$$

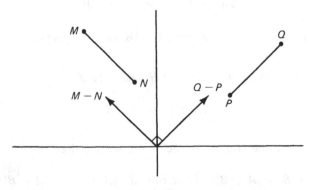

Figure 10.10

Observe that the order in which we take P, Q or M, N is unimportant, and that we could have defined perpendicularity just as well for segments. Also, we could say according to our usual practice that a segment $\overrightarrow{PQ}$ is perpendicular to A if

$$(P - Q) \cdot A = 0.$$

In this context, we visualize A as the endpoint of the located vector $\overrightarrow{OA}$.

Example. Let $P = (3, 4)$, $Q = (-5, 1)$, $M = (7, 3)$, and $N = (4, 11)$. Then $\overrightarrow{PQ}$ is perpendicular to $\overrightarrow{MN}$. To verify this, note that

$$Q - P = (-8, -3) \quad \text{and} \quad N - M = (-3, 8).$$

Hence

$$(Q - P) \cdot (N - M) = (-8)(-3) + (-3)8 = 0.$$

This proves our assertion.

Example. Let $P = (-1, 2)$, $Q = (3, 1)$, and $M = (-2, -4)$. Find a point N such that $\overrightarrow{MN}$ is perpendicular to $\overrightarrow{PQ}$, and such that the length of $\overrightarrow{MN}$ is equal to the length of $\overrightarrow{PQ}$.

This is easily done. First we compute

$$Q - P = (4, -1).$$

Then $(1, 4)$ is perpendicular to $(4, -1)$ (their dot product is 0). Let

$$N = M + (1, 4) = (-2, -4) + (1, 4) = (-1, 0).$$

Then $N - M = (1, 4)$ which has been selected such that it is perpendicular to $Q - P$. Furthermore, $(1, 4)$ has the same norm as $(4, -1)$, because

$$\sqrt{1^2 + 4^2} = \sqrt{4^2 + (-1)^2}.$$

Thus our choice $N = (-1, 0)$ solves our problem.

We draw a generalized picture below.

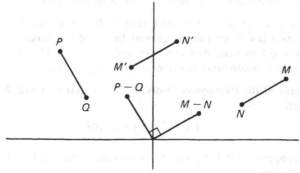

Figure 10.11

It is now easy to prove the following statement:

If A is perpendicular to B and C = xB for some number x, then A is perpendicular to C.

We leave the proof as Exercise 13(b).

10, §3. EXERCISES

In each of the following cases, determine whether A is perpendicular to B.

1. $A = (4, 5)$, $B = (-3, 1)$ 2. $A = (4, 5)$, $B = (-5, 4)$

3. $A = (4, 5)$, $B = (10, -8)$ 4. $A = (-2, 7)$, $B = (-14, 4)$

5. $A = (-2, 7)$, $\mathbf{B} = (-14, -4)$ 6. $A = (-2, 7)$, $B = (14, -4)$

In each of the following cases, determine whether $\overline{PQ}$ is perpendicular to $\overline{MN}$.

7. $P = (1, 5)$, $Q = (3, 6)$, $M = (7, 2)$, $N = (11, 4)$

8. $P = (4, -6)$, $Q = (5, -1)$, $M = (6, 1)$, $N = (-1, 4)$

In each of the following cases, determine two points N such that $\overline{MN}$ is perpendicular to $\overline{PQ}$, and such that the length of $\overline{MN}$ is equal to the length of $\overline{PQ}$.

9. $P = (1, 5)$, $Q = (3, 6)$, $M = (7, 2)$

10. $P = (-3, 4)$, $Q = (-2, -1)$, $M = (-4, -5)$

11. $P = (4, -1)$, $Q = (5, 3)$, $M = (7, -4)$

12. $P = (3, 1)$, $Q = (-1, 1)$, $M = (2, -1)$

13. Prove:
    (a) If A is perpendicular to B and A is perpendicular to C, then A is perpendicular to $B + C$.
    (b) If A is perpendicular to B and $C = xB$ for some number x, then A is perpendicular to C.
    Use only properties **SP 1**, **SP 2** and **SP 3** in your proofs.

14. (a) Give an example of A, B, C such that $A \cdot B = A \cdot C$ but $B \neq C$. **Thus the cancellation law is not valid in general for the dot product**.
    (b) If $A \cdot X = 0$ how does $A \cdot B$ compare with $A \cdot (B + X)$? How does this remark help you construct many examples in part (a)?

15. **Analytic form of the Pythagoras theorem**. Prove that if A and B are perpendicular, then

$$|A + B|^2 = |A|^2 + |B|^2.$$

Use only properties **SP 1** through **SP 4** in your proof, and also in the proof of the next exercises.

16. (a) Prove: For any A, B we have:

$$|A + B|^2 + |A - B|^2 = 2|A|^2 + 2|B|^2.$$

   (b) Draw a parallelogram with corners at O, A, B, $A + B$. How does the re-
       lation of (a) establish a relation between the lengths of the sides of the
       parallelogram and the lengths of the diagonals?

17. Prove that for any A, B we have

$$|A + B|^2 - |A - B|^2 = 4A \cdot B.$$

18. Let A, B be perpendicular and not equal to O. Let x, y be numbers such
    that

$$xA + yB = O.$$

Prove that $x = y = 0$. [*Hint*: Dot both sides with A and B respectively.]

10, §4. PROJECTIONS

Let us return to the airplane and the wind. Suppose the airplane wants
to go in a direction represented by a vector B, and the wind blows in a
direction shown by the vector W in the figure.

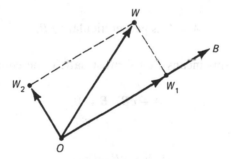

Figure 10.12

We can decompose W into two parts, one which points in the same
direction as B, and one which points into a perpendicular direction.
These are denoted by W_1 and W_2 respectively, so that

$$W = W_1 + W_2 \quad \text{and} \quad W_1 \perp W_2.$$

The directions are determined from the origin O. The part W_1 as drawn
in the figure would push the plane in the direction it wants to go. The

part W_2 would push the plane in a perpendicular direction, where the plane does not want to go. We shall now study how to compute those parts W_1 and W_2 algebraically, which would be needed by a navigator trying to direct the course of the plane.

Let A, B be points, and assume $B \neq O$. Let P be the point on the line through O, B such that $\overrightarrow{PA}$ is perpendicular to $\overrightarrow{OB}$, as shown on Figure 10.13(a).

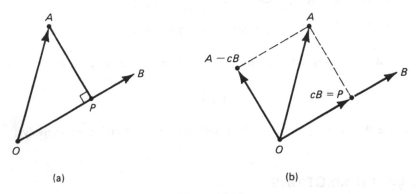

(a)   (b)

Figure 10.13

We can write $P = cB$ for some number c. We want to find this number c explicitly in terms of A and B. The condition $\overrightarrow{PA} \perp \overrightarrow{OB}$ means that

$$A - P \text{ is perpendicular to } B,$$

and since $P = cB$ this means that c must satisfy the equation

$$(A - cB) \cdot B = 0,$$

or in other words,

$$A \cdot B - cB \cdot B = 0.$$

We can solve for c, and we find $A \cdot B = cB \cdot B$, so that

$$\boxed{c = \frac{A \cdot B}{B \cdot B}.}$$

Conversely, take this value for c, and use distributivity. Then you find that $(A - cB) \cdot B = 0$, so that $A - cB$ is perpendicular to B. Hence we have seen that there is a unique number c such that $A - cB$ is perpendicular to B. We call this number c the **component of A along B** and we call cB the **projection of A along B**.

Example. Let $A = (3, 1)$ and $B = (5, -2)$. Then the **component** of A along B is the number

$$c = \frac{A \cdot B}{B \cdot B} = \frac{15 - 2}{25 + 4} = \frac{13}{29}.$$

The **projection** of A along B is the point cB, namely

$$cB = \frac{13}{29} (5, -2) = \left(\frac{65}{29}, \frac{-26}{29} \right).$$

Example. Let

$$E_1 = (1, 0) \quad \text{and} \quad E_2 = (0, 1).$$

We shall call E_1 and E_2 the **basic unit points** in $\mathbf{R}^2$. Let $A = (a_1, a_2)$. Then we note that $A = a_1 E_1 + a_2 E_2$. Furthermore,

$$A \cdot E_1 = a_1 \quad \text{and} \quad A \cdot E_2 = a_2.$$

Thus the component of A along E_1 is just the first coordinate of A, and the component of A along E_2 is the second coordinate of A.

Example. Suppose that B has norm 1, so $|B| = 1$. Then $B \cdot B = 1$. For an arbitrary point A, we see that the component of A along B is simply $A \cdot B$. A point B such that $|B| = 1$ is called a **unit point**. We see that the dot product of A with a unit point is a generalization of the dot product of A with the basic unit points of the preceding example. We also have a geometric interpretation of the dot product $A \cdot B$ as giving a component of A along B, if B is a unit point.

10, §4. EXERCISES

1. Find the component of A along B in each of the following cases.
   (a) $B = (5, 7)$ and $A = (3, -4)$
   (b) $B = (2, -1)$ and $A = (-3, -4)$
   (c) $B = (-3, -2)$ and $A = (5, 1)$
   (d) $B = (-4, 1)$ and $A = (2, -5)$

2. Find the projection of A along B in each one of the cases of Exercise 1.

10, §5. ORDINARY EQUATION FOR A LINE

The theory of the dot product and perpendicularity give us a way of writing down the equation of a straight line in the plane by one equation. We want to find a simple condition for a point $X = (x, y)$ to lie on the line L passing through a given point P, and perpendicular to a given vector $\overrightarrow{ON}$, with $N \neq O$. This is easily done. A point X lies on the line if and only if the segment $\overline{PX}$ is perpendicular to $\overrightarrow{ON}$. Going back to the definition of §3 concerning the perpendicularity of such segments, we see that X lies on the line if and only if

(∗) $$(X - P) \cdot N = 0.$$

This is illustrated by Figure 10.14.

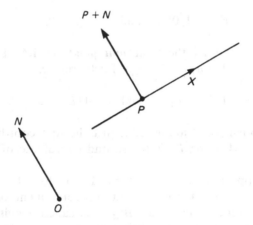

Figure 10.14

Using the distributivity of the dot product, we can rewrite equation (∗) in the form

$$X \cdot N - P \cdot N = 0,$$

or also

$$X \cdot N = P \cdot N.$$

These equations are called the **ordinary equations** for a line passing through P, perpendicular to N.

Example. Let $P = (2, 5)$ and $N = (-3, 7)$. We wish to find the ordinary equation for the line through P perpendicular to N. Let $X = (x, y)$ denote a point of the plane. Then X lies on this line if and only if

$$X \cdot N = P \cdot N.$$

With our special numbers, this yields

$$-3x + 7y = 2(-3) + 5 \cdot 7 = 29;$$

that is,

$$-3x + 7y = 29.$$

Example. Let $P = (-4, 1)$. Find the ordinary equation of the line passing through P and perpendicular to the segment $\overline{QM}$, where

$$Q = (3, 2)$$

and

$$M = (-1, 4).$$

Let $N = M - Q = (-4, 2)$. We must find the ordinary equation of the line passing through P and perpendicular to N. It is

$$-4x + 2y = P \cdot N = 16 + 2 = 18;$$

that is,

$$-4x + 2y = 18.$$

In general, let P be a given point. Let $N = (a, b)$, $N \neq O$. Let

$$P \cdot N = c.$$

Then the equation of the line passing through P and perpendicular to N is

$$\boxed{ax + by = c.}$$

Example. The equation

$$5x - 7y = 9$$

is the equation of a line perpendicular to $N = (5, -7)$. It is easy to find a point on this line. For instance, take an arbitrary x-coordinate, say $x = 2$, and solve for y. We get

$$y = \frac{9 - 10}{7} = \frac{1}{7}.$$

Thus $P = (2, \frac{1}{7})$ is a point on this line. Note how $P \cdot N = 9$, as it should.

Let L be a line perpendicular to N, and let L' be a line perpendicular to N'. We can take it as a definition that L **is perpendicular to** L' if and only if N is perpendicular to N'. It is now easy to recognize from the equations of two lines whether they are perpendicular.

Example. Let L be the line defined by the equation

$$5x - 7y = 9,$$

and let L' be the line defined by the equation

$$3x + 4y = 2.$$

Then L is perpendicular to $N = (5, -7)$ and L' is perpendicular to

$$N' = (3, 4).$$

Since

$$N \cdot N' = 15 - 28 = -13 \neq 0,$$

we conclude that L is not perpendicular to L'.

Example. Let L be the line defined by the equation

$$5x - 7y = 9,$$

and let L' be the line defined by the equation

$$14x + 10y = 17.$$

Then L is perpendicular to $N = (5, -7)$, and L' is perpendicular to

$$N' = (14, 10).$$

Since

$$N \cdot N' = 5(14) - 7(10) = 70 - 70 = 0,$$

we see that N is perpendicular to N', whence L is perpendicular to L'.

If a line L is defined by an equation of the form

$$y = mx + b,$$

recall that m is called the **slope** of the line. We can now give a condition on the slopes of two lines for them to be perpendicular.

Theorem 10-3. *Let $y = m_1 x + b_1$ and $y = m_2 x + b_2$ be the equations of two lines. They are perpendicular if and only if*

$$m_1 m_2 = -1.$$

Proof. The equations for the lines can be written in the form

$$m_1 x - y = -b_1 \quad \text{and} \quad m_2 x - y = -b_2.$$

Let $N = (m_1, -1)$ and $N' = (m_2, -1)$. Then N, N' are perpendicular vectors to the lines, and the condition of perpendicularity for the lines is precisely that the dot product of N and N' is 0, that is

$$0 = N \cdot N' = m_1 m_2 + 1.$$

This is equivalent with $m_1 m_2 = -1$, as desired.

Example. The lines $y = 3x - 5$ and $y = 4x + 1$ are not perpendicular because $3 \cdot 4 = 12 \neq -1$.

The lines $y - 2x + 9$ and $y = -\frac{1}{2}x + 6$ are perpendicular because

$$2 \cdot (-\tfrac{1}{2}) = -1.$$

10, §5. EXERCISES

Find the ordinary equation of the line perpendicular to N and passing through P for the following values of N and P.

1. (a) $N = (1, -1)$, $P = (-5, 3)$ 2. (a) $N = (-5, 4)$, $P = (3, 2)$
   (b) $N = (-3, 5)$, $P = (7, -2)$ (b) $N = (2, 8)$, $P = (1, 1)$

3. (a) $N = (4, 1)$, $P = (5, -2)$
   (b) $N = (-5, 3)$, $P = (6, -7)$

4. Are the lines with equations

$$3x - 5y = 1, \qquad 2x + 3y = 5$$

perpendicular?

5. Which of the following pairs of lines are perpendicular?
   (a) $3x - 5y = 1$ and $2x + y = 2$
   (b) $2x + 7y = 1$ and $x - y = 5$
   (c) $3x - 5y = 1$ and $5x + 3y = 7$
   (d) $-x + y = 2$ and $x + y = 0$

6. Which of the following pairs of lines are perpendicular?
   (a) $y = -x + 4$ and $y = 3x + 1$
   (b) $y = 3x - 1$ and $y = -\frac{1}{3}x + 8$
   (c) $y = 4x - 5$ and $y = -\frac{1}{4}x - 7$
   (d) $y = \frac{1}{2}x + 1$ and $y = 2x - 3$

10, §6. THE 3-DIMENSIONAL CASE

We assume that you have done previous exercises on points in 3-space, and that you know about addition of such points, and their multiplication by numbers. The properties are entirely similar to those of the 2-dimensional case.

We can then define the dot product in 3-space in a manner completely similar to the dot product in 2-space. Indeed, if

$$A = (a_1, a_2, a_3) \qquad \text{and} \qquad B = (b_1, b_2, b_3),$$

then we define their **dot** (or **scalar**) **product** to be the number

$$A \cdot B = a_1 b_1 + a_2 b_2 + a_3 b_3.$$

Example. The dot product of $A = (3, -1, 4)$ and $B = (-2, 5, 3)$ is

$$A \cdot B = 3(-2) + (-1)5 + 4(3)$$
$$= -6 - 5 + 12 = 1.$$

The dot product satisfies the same properties **SP 1** through **SP 4** which we stated before. The proofs for these properties must use now three coordinates instead of two, but otherwise, they are entirely analogous, and we can leave them to you. The idea is that the proofs reduce these properties about *vectors* with *coordinates* to similar properties about *numbers*. Just to convince you how easy they are, we shall give the proof of **SP 3**.

Let x be a number. Let $A = (a_1, a_2, a_3)$ and $B = (b_1, b_2, b_3)$. Then $xA = (xa_1, xa_2, xa_3)$ by the definition of the product of a number times a point. Again by the definition of the dot product, we have:

$$(xA) \cdot B = xa_1 b_1 + xa_2 b_2 + xa_3 b_3.$$

We can now factor out x from this expression, which we see is equal to

$$x(a_1 b_1 + a_2 b_2 + a_3 b_3) = x(A \cdot B).$$

This proves that $(xA) \cdot B = x(A \cdot B)$, as we wanted.

You will find that your proofs for the formulas giving the expansion of $(A + B)^2$, $(A - B)^2$, and $(A + B) \cdot (A - B)$ also work in the present case, because they should have used only **SP 1** and **SP 2**.

We can also define the **norm** of A to be

$$|A| = \sqrt{a_1^2 + a_2^2 + a_3^2}.$$

In other words,

$$A \cdot A = |A|^2 \quad \text{and} \quad |A| = \sqrt{A \cdot A}.$$

Example. Let $A = (-4, 2, 3)$. Then $|A| = \sqrt{16 + 4 + 9} = \sqrt{29}$. Observe that the norm of A gives the distance $d(O, A)$ from the origin to the point A, using Pythagoras' theorem illustrated in Figure 10.15. This was discussed in Chapter 4, §2.

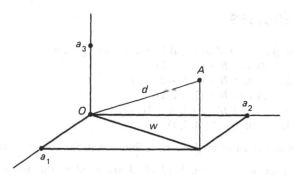

Figure 10.15

As in the 2-dimensional case, we define A to be **perpendicular** to B if and only if $A \cdot B = 0$. The argument of Theorem 10-2 applies in the present case, because it used only properties **SP 1** through **SP 4**, which are valid. Thus Theorem 10-2 justifies our definition psychologically. Having this definition of perpendicularity, we can now define the **component** of A along B as before, to be the number

$$c = \frac{A \cdot B}{B \cdot B},$$

provided of course that $B \neq O$. The **projection** of A along B is defined to be cB.

Example. Let $A = (-3, 1, 4)$ and $B = (2, 1, -1)$. Then the *component* of A along B is the *number*

$$\frac{A \cdot B}{B \cdot B} = \frac{-6 + 1 - 4}{4 + 1 + 1} = \frac{-9}{6} = \frac{-3}{2}.$$

The *projection* of A along B in the present case is therefore equal to

$$cB = (-3, -\tfrac{3}{2}, \tfrac{3}{2}).$$

To find the distance $d(A, B)$ we compute $A - B$ which is

$$A - B = (-5, 0, 5).$$

Hence

$$d(A, B)^2 = 25 + 0 + 25 = 50,$$

and $d(A, B) = \sqrt{50}$.

10, §6. EXERCISES

1. Find the component of A along B in the following cases.
   (a) $A = (3, -1, 4)$ and $B = (1, 2, -1)$
   (b) $A = (1, 2, -1)$ and $B = (-1, -2, -5)$
   (c) $A = (3, -1, 2)$ and $B = (2, -2, 1)$
   (d) $A = (-4, 3, 1)$ and $B = (-1, 3, -2)$
   (e) $A = (1, 1, 1)$ and $B = (-4, 2, 1)$

2. In each one of the cases of Exercise 1, find the projection of A along B.

3. Verify that Exercises 13 through 17 of §3 are valid in the 3-dimensional case. In fact, their proofs should have used only **SP 1** through **SP 4**, which shows that they are valid.

4. Find the norm of A in each one of the cases of Exercise 1.

5. Find the distance between A and B in each one of the cases of Exercise 1.

6. Let A, B, C be non-zero elements of $\mathbf{R}^3$, which are mutually perpendicular (that is, any two of them are perpendicular to each other). If x, y, z are numbers such that

$$xA + yB + zC = 0,$$

prove that $x = y = z = 0$.

7. Let $E_1 = (1, 0, 0)$, $E_2 = (0, 1, 0)$, and $E_3 = (0, 0, 1)$. We call these the **basic unit points** in $\mathbf{R}^3$. If $A = (a_1, a_2, a_3)$, find the dot product of A with E_1, E_2, E_3 respectively.

8. Define a point in $\mathbf{R}^4$ (4-dimensional space) to be a quadruple of numbers $A = (a_1, a_2, a_3, a_4)$. For instance,

$$(-1, 3, 5, 2)$$

is a point of 4-space. Define the addition of such points, and multiplication by numbers. Define the dot product in a manner similar to that of the text, and show that **SP 1** through **SP 4** are valid. How about in n-dimensional space?

10, §7. EQUATION FOR A PLANE IN 3-SPACE

The discussion of §5 concerning the equation of a line in 2-space generalizes to that of a plane in 3-space as follows. Given a point $N \neq O$, which we visualize as the endpoint of the located vector $\overrightarrow{ON}$, we wish to write down an equation for all points lying in the plane passing through a given point P, and **perpendicular** to $\overrightarrow{ON}$ (or as we also say, **perpendicular** to N). Again, we see from Figure 10.16 what condition a point X must satisfy in order to lie on the plane, namely $\overrightarrow{PX}$ must be perpendicular to $\overrightarrow{ON}$. This is true, by definition, if and only if

$$(X - P) \cdot N = 0,$$

or in other words,

$$\boxed{X \cdot N = P \cdot N,}$$

This is exactly the same formula as in 2-space, but it now involves three coordinates of a point $X = (x, y, z)$.

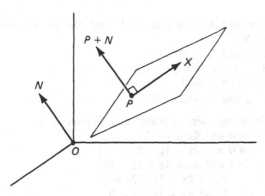

Figure 10.16

Example. Let $X = (x, y, z)$. Let $N = (3, -2, 4)$. Find the equation of the plane passing through the point $P = (-1, 3, 5)$, and perpendicular to N. This is easily done. We have

$$P \cdot N = -3 - 6 + 20 = 11,$$

so that the equation of the desired plane is

$$3x - 2y + 4z = 11.$$

Example. The equation

$$7x - 5y - 8z = 33$$

is the equation of a plane perpendicular to $N = (7, -5, -8)$.

We can also determine when two planes are perpendicular, using the same criterion as for lines.

Example. Let two planes be defined by the equations

$$3x - 5y + 4z = 2 \quad \text{and} \quad -4x + 7y - z = 1.$$

Then the first plane is perpendicular to $N = (3, -5, 4)$ and the second plane is perpendicular to $N' = (-4, 7, -1)$. Since

$$N \cdot N' = -12 - 35 - 4 \neq 0,$$

it follows that the planes are not perpendicular.

10, §7. EXERCISES

1. Find the equation of the plane passing through the given point P, perpendicular to the given N.
   (a) $P = (-4, 5, 1)$ and $N = (2, 3, 1)$
   (b) $P = (1, -3, -7)$ and $N = (-2, 1, 4)$
   (c) $P = (-2, 1, 3)$ and $N = (1, -1, 1)$
   (d) $P = (3, 1, -1)$ and $N = (2, 1, 3)$

2. Which of the following pairs of planes are perpendicular?
   (a) $3x - 2y + 5z = 0$ and $4x - y + 7z = 1$
   (b) $3x - 2y + 5z = 0$ and $7x - 3y + 2z = -1$
   (c) $3x - 2y - 5z = 8$ and $6x + 4y + 2z = 17$
   (d) $3x + 4y + 11z = 2$ and $2x - 7y + 2z = 21$
   (e) $7x - y + 8z = -2$ and $8x + 2y - z = 0$

CHAPTER 11

Transformations

11, §1. INTRODUCTION

Until now, we have studied primarily the properties of geometric figures. Now we will begin dealing with *relationships* between figures. We pursue the discussion which began Chapter 7. Look at the three triangles in Figure 11.1.

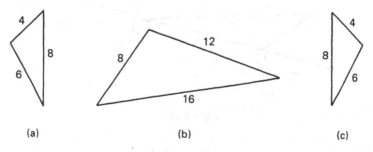

Figure 11.1

Triangles (a) and (c) have the same dimensions, but we had already noted in Chapter 7 that they are not "the same". On the other hand, the lengths of the sides of Triangle (b) are twice the lengths of the sides of Triangle (a). In what other ways are these two triangles related? What about their areas, or the sizes of their angles?

In Figure 11.2, two clay pots are illustrated:

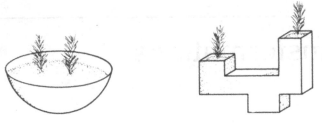

Figure 11.2

The one on the left seems more "regular" than the one on the right. What property (geometrically speaking) does it have, which the other one doesn't?

Print the word MOM on a piece of paper using capital letters, and then hold it up to a mirror. Can you read the word in the mirror? Try it again with the word RED. What happens this time? What is different between MOM and RED? How is this all related to clay pots, for instance?

Finally, consider this problem. A manufacturing company has two warehouses (A and B) located along a river bank as illustrated:

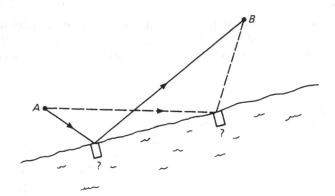

Figure 11.3

The company wishes to build a pier on the river near the warehouses so that supply boats can dock. A truck will deliver finished goods to the pier for loading onto the boat, pick up raw materials, and drive them to warehouse B for storage. Where should the company locate the pier so that the distance from warehouse A (where the truck starts) to the pier, plus the distance from the pier to warehouse B is as short as possible?

This would make the driver's trip each time as short as possible, saving the company money.

We will be able to answer these questions and many more using the notion of a geometric mapping.

11, §1. EXERCISES

1. Illustrated below are two right triangles with measurements given:

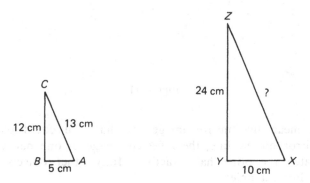

Figure 11.4

(a) Can you find length $|XZ|$ from the information given?
(b) How are the lengths of the sides of these triangles related?
(c) How are the perimeters related?
(d) Find the areas of the triangles. How are they related?
(e) Can you find any other relationships between the triangles?

2. Do the experiment with the word MOM and a mirror.
   (a) Find two other words besides MOM which have the same property.
   (b) Describe as best you can, using geometric ideas, why these words "work" the way they do in a mirror.

EXPERIMENT 11-1

Work on the warehouse problem. Draw a similar diagram on a piece of paper. Pick a possible location for the pier, and carefully measure the distances involved. Compare notes with your classmates. When you think you have a method for determining the solution, see if you can prove that your method does indeed give the shortest path. [*Hint*: The easiest solution involves line symmetry.]

11, §2. SYMMETRY AND REFLECTIONS

If you draw a line down the middle of the human body, the two halves "match" each other. One half is the "mirror image" of the other, as illustrated in Figure 11.5.

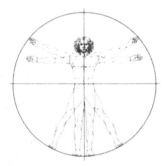

Figure 11.5

What we mean by "mirror image" is that you could place a small pocket mirror on the line; the reflected image of one half of the figure would match the other half exactly. Many other objects exhibit this property. For example:

Figure 11.6

You can probably suggest many more man-made or natural objects which can be divided into matched halves. Naturally, we would like to have a good mathematical description of this property for figures in the plane. A good set of figures to start investigating are the letters of the alphabet, written in block capitals: A, B, C, D, E, F, G, H, I, etc.

Consider the letter H. If we draw a line down through its middle, we divide it into halves which are mirror images of each other:

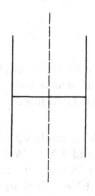

Figure 11.7

We can also draw a horizontal line through the H which also divides it into mirror image halves (up and down, rather than left and right):

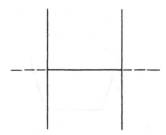

Figure 11.8

The letter E only has a horizontal dividing line:

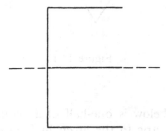

Figure 11.9

Such lines are called **lines of symmetry**. A figure which has one or more lines of symmetry is said to have **line symmetry**. Thus the letter H has line symmetry, as does a plane representation of the human body, the Eiffel tower, our Capitol, etc. We also say that these figures are **symmetrical**.

EXPERIMENT 11-2

1. Find lines of symmetry for as many capital letters as you can. Are there other letters besides H which have more than one line of symmetry? Are there any which have none?

2. Notice that the word MOM is made up of letters which are symmetrical. So is the word TOO and EYE. Write TOO on a piece of paper and hold it up to a mirror. Can you read it in the mirror? Do the same thing with the word EYE. Explain the differences in the words MOM, TOO, and EYE.

3. Below are some geometric shapes. Find as many lines of symmetry for each of them as you can.

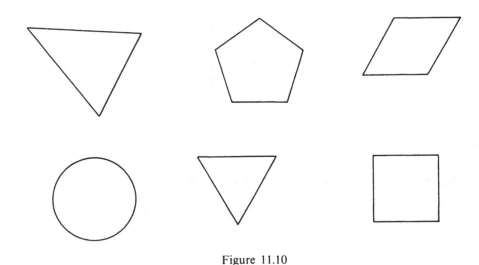

Figure 11.10

4. Each of the shapes below is one-half of a geometric figure; the other missing half is symmetric to the given half (in other words, the given line is a line of symmetry). Draw the missing half of each figure.

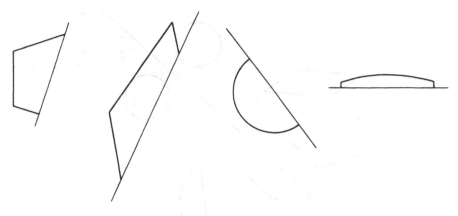

Figure 11.11

5. Try to state in mathematical terms what the property of line symmetry is. Your description should use the concepts of point, line, distance, etc., and not use phrases such as "if you held up a mirror ...", etc.

From your work with the Experiment, you should see that a line of symmetry divides a figure into two halves. If we consider the symmetry line as a mirror, it's as if each half of the figure were a "reflection" of the other half. Of course, not all geometric figures have symmetry lines.

This idea of reflection is useful in relating two separate figures, as well as giving us the idea of symmetry in a single figure. Look at Figure 11.12; do you see that triangle $A'B'C'$ is the "reflection" of triangle ABC through line L?

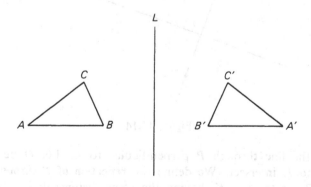

Figure 11.12

Line L is acting like a symmetry line for the whole plane! Figure 11.13 shows a line, three figures, and their reflections through the line.

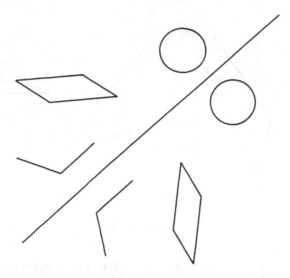

Figure 11.13

By giving the idea of reflection through a line a mathematical defini-tion, we will at the same time be able to define exactly what symmetry is. You may already have a good definition for symmetry as your answer to Exercise 5 in the Experiment.

To define reflection, we must start with a line. We wish to define the reflection of a point through this line. So ... given line L, and any point P:

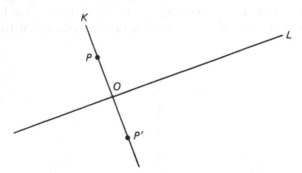

Figure 11.14

Let K be the line through P perpendicular to L. Let O be the point where L and K intersect. We define the **reflection of P through L** to be the point P' on the line K, having the same distance from O as P, but on the other side of L. Mathematicians borrow a word from optics, and say that P' is the **image** of P, just as we talk about the image in a mirror. If P happens to lie on line L, then P' is just point P itself.

To reflect a whole figure through a line, we just reflect all of its points. Looking back at Figure 11.12, visualize reflecting each point in triangle *ABC* through line *L* to get a new triangle, *A'B'C'*. As you look at the figure, you might ask: How do we know that all the points on line segment $\overline{AB}$, for instance, will reflect into another line segment? Maybe the reflection of a line segment is not straight, but a curve, as shown:

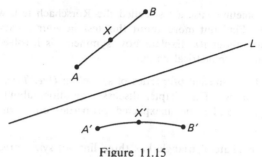

Figure 11.15

For the moment, this certainly doesn't seem likely from our definition, and we will see later that it is not so.

Finally, we can define line symmetry, as we set out to do at the beginning. If you can draw a line *L* so that the reflection of the figure through this line coincides with the figure itself, we say that the figure has **line symmetry**.

CONSTRUCTION 11-1

Constructing the reflection of a point through a given line.

Given line *L* and some point *P*. We wish to construct the reflection of point *P* through *L*.

First, construct a line through *P* perpendicular to line *L* (use Construction 5-2). Label this line *K*, and label the intersection of *K* and *L* point *O*:

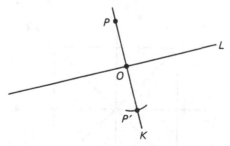

Figure 11.16

Set the tips of your compass at a distance equal to $|PO|$. With the tip of the compass on *O*, draw an arc across line *K* on the opposite side of *O* from *P*. This arc crosses line *K* at a point *P'*, which is the **reflection** of point *P*.

11, §2. EXERCISES

1. What symmetry properties does the phrase WOW MOM WOW have? Can you think of other phrases which have all these properties? Look up the meaning of the word *palindrome* as a clue.

2. Psychologists sometimes use a test called the Rorschach test, which is a set of inkblot designs. Find out more about this test in your library (encyclopedias have information about it). Explain how symmetry is involved, and why you think the test uses symmetrical designs.

3. Read Chapter III, "Unusual properties of space" in *One, Two, Three ... Infinity* by George Gamow. This chapter discusses questions about symmetry in 3-dimensional space, and some unexpected properties our own universe might have.

4. Notice that an equilateral triangle has three lines of symmetry:

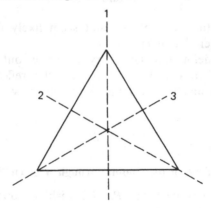

Figure 11.17

and that a square has four:

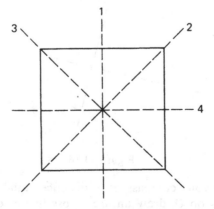

Figure 11.18

(a) How many lines of symmetry does a regular 5-gon (pentagon) have?

(b) How many lines of symmetry does a regular 6-gon (hexagon) have?

(c) Draw a 4-gon which has *no* lines of symmetry.

11, §3. TERMINOLOGY

Reflection through a line is an example of what is called a "geometric mapping". This idea of mapping is a very basic one in mathematics, and with it we will be able to answer all the questions posed in the "Introduction" section, as well as many others.

Recall the definition of reflection. Given a line L in the plane, we can associate with any point P in the plane another point P', which is its reflection through L. Thus the rule of reflection gives us a way of "associating" points—each point in the plane gets associated with another point in the plane which we have called its image. The definition of mapping uses precisely this idea:

A **mapping** (or **map**) of the plane into itself is an association, which to each point of the plane associates another point of the plane.

We say "... of the plane into itself" to mean that each point in the plane has an associate (no one is "left out"), and that this associated point also lies in the plane (there are mappings which associate points in the plane with points in space).

If P is a point, and P' is the point associated with P by the mapping, we denote this by the special arrow:

$$P \mapsto P'.$$

We say that P' is the **image** of P (even if the mapping is something other than reflection), or that P is **mapped** onto P'.

Just as we sometimes use letters to stand for numbers, it is useful to use letters to denote mappings. For example, we sometimes say "let x be a number" when we want to prove something about numbers. Similarly, we will say "let F be a mapping" when F stands for a mapping of the plane (like reflection, for example).

If F is a mapping, we denote the image of a point P by the symbols:

$$F(P) \qquad \text{(read ``}F \text{ of } P\text{'')}.$$

If we label this image point P', we can then write:

$$P' = F(P).$$

This equation tells us that the image of P according to the mapping F is the point called P'. Another way of saying this is that F **maps** P **onto** P', or that P' is the **value** of F at P.

Example. Let F be reflection through line L, and let X be some point in the plane, as illustrated:

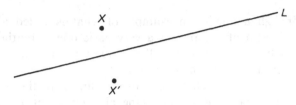

Figure 11.19

In this situation, the expressions

$$X \mapsto X'$$

or

$$F(X) = X'$$

or

$$X' \text{ is the } image \text{ of } X \text{ under } F$$

or

$$X' \text{ is the } value \text{ of } F \text{ at } X$$

mean the same thing.

It turns out that certain mappings are more interesting and useful than others. In the remaining sections of this chapter we describe these standard mappings.

11, §4. REFLECTION THROUGH A LINE

This is a mapping we have encountered already. Let L be a line. If P is any point in the plane, let line K be the line through P perpendicular to line L. Let O be the point of intersection of L and K. Let P' be the point on K which is at the same distance from O as P, but in the opposite direction. See Figure 11.20. Then P' is the image of P under this mapping, which is called **reflection through line L**.

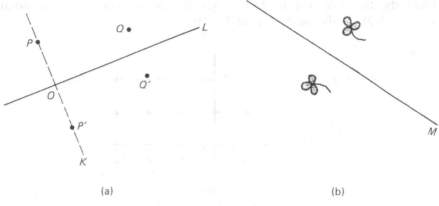

(a)   (b)

Figure 11.20

We denote this mapping with a special abbreviation R_L. We would then write:

$$\boxed{P' = R_L(P)}$$ (read "R sub L of P").

In Figure 11.20(b), we show the reflection of a flower through a line M.

We now describe reflection through a line in terms of coordinates. We shall do this in the important special cases when the line is one of the coordinate axes. For instance, let us start with the line being the x-axis. Let $P = (3, 2)$. What are the coordinates of the point P' which is the reflection of P through the x-axis? We see that $P' = (3, -2)$ as shown on Figure 11.21(a).

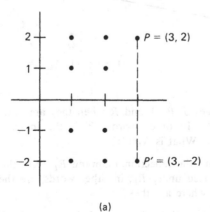

(a)

Figure 11.21(a)

Similarly, the reflection of P through the y-axis would be the point $P'' = (-3, 2)$, as shown on Figure 11.21(b).

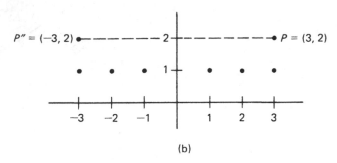

(b)

Figure 11.21(b)

In general, if $P = (x, y)$ then:

$(x, -y)$ is the reflection of P through the x-axis,

$(-x, y)$ is the reflection of P through the y-axis.

11, §4. EXERCISES

1. (a) Copy line L and points P, Q, and R on your paper.

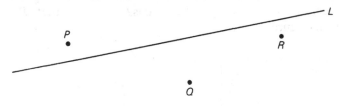

Figure 11.22

Draw the images of P, Q, and R when they are reflected through line L.
(b) Let $P' = R_L(P)$. In other words, P' is the image of P when reflected through line L. What is $R_L(P')$?

2. Given a line L, and the reflection mapping R_L. Are there any points in the plane which stay fixed under R_L; in other words, are there any points where $R_L(P) = P$? If so, where are they?

3. Redefine the reflection through a line L by using the idea of perpendicular bisector.

4. In Figure 11.23, P' is the image of P when reflected through L, and Q is on line L.   Prove that $d(P, Q) = d(P', Q)$.

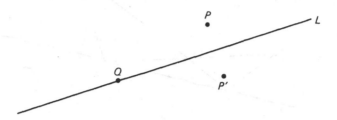

Figure 11.23

5. In the diagram below, $X' = R_L(X)$. Construct, using a *straightedge only* (a ruler without any numbers), the point $R_L(Y)$.

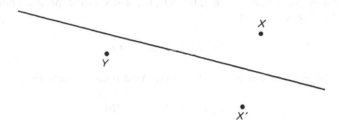

Figure 11.24

6. For each point P, write the coordinates of its image P' when reflected through the x-axis:
   (a) $P = (3, 7)$ (b) $P = (-2, 5)$ (c) $P = (5, -7)$
   (d) $P = (-5, -7)$ (e) $P = (p_1, p_2)$

7. For each point P above, write the coordinates of its image p'' when reflected through the y-axis.

8. What is the image of the line $x = 3$ when reflected through the x-axis? In other words, if we reflect every point on the line $x = 3$ through the x-axis, what figure would result?

9. What is the image of the line $x = 3$ when reflected through the y-axis?

10. The result of Exercise 4 can be used to solve the problem about the two warehouses given at the beginning of this chapter. You want to find the point X on the river bank so that the distance $|AX|$ plus the distance $|XB|$ is as short as possible.

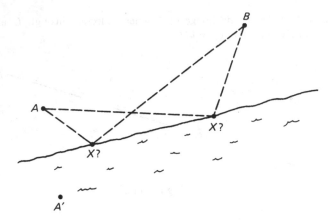

Figure 11.25

Reflect point A through the line determined by the river bank to get point A'. Where line segment $\overline{A'B}$ intersects the bank is the proper point X. Use Exercise 4 to show that

$$|AX| + |XB| = |A'B|.$$

Let Y be any other point on the bank. You need to show that

$$|AY| + |YB| > |A'B|.$$

Use Exercise 4 again and the triangle inequality. This shows that any other location for the pier besides point X will cause the truck driver's route to be longer.

11. Let R_L be the reflection through line L. Let P and Q be two points in the plane, and let $P' = R_L(P)$ and $Q' = R_L(Q)$. Prove that

$$d(P, Q) = d(P', Q').$$

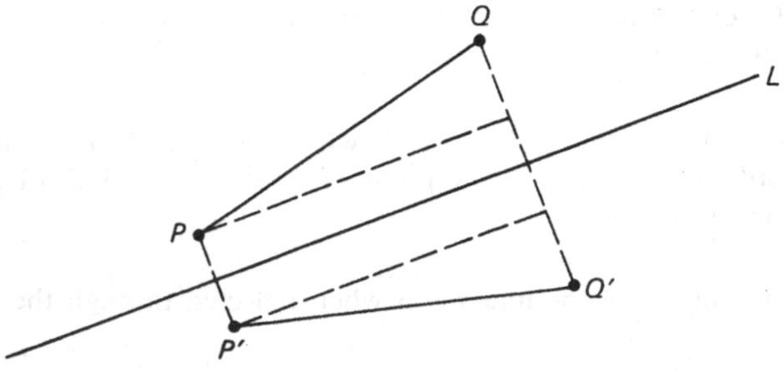

Figure 11.26

[*Hint*: Draw lines like those indicated.]

12. Exercise in 3-space

Let P be a point in 3-space. Can you visualize how it could be reflected through the (x, y)-plane? Suppose P had coordinates (p_1, p_2, p_3). What would the coordinates of its image be?

Suppose we reflected P through the (x, z)-plane. What would the coordinates of its image be? Same question for reflection through the (y, z)-plane.

11, §5. REFLECTION THROUGH A POINT

Let O be a given point of the plane. To each point P of the plane, we associate the point P' lying on the line passing through P and O, on the other side of O from P, and at the same distance from O as P. See Figure 11.27(a).

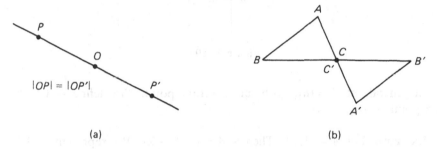

(a)   (b)

Figure 11.27

This mapping is called **reflection through point** O. In Figure 11.27(b), we show a triangle reflected through one of its vertices, $C = C'$.

Suppose that O is the origin of our coordinate axes. We want to describe reflection through O in terms of coordinates.

First, consider a point $P = (3, 0)$ on the x-axis. We see that its reflection through the origin is the point $(-3, 0)$ as shown on Figure 11.28.

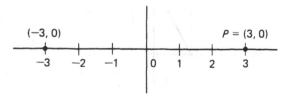

Figure 11.28

Second, consider a point $(0, 3)$ on the y-axis. Its reflection through 0 is the point $(0, -3)$ as shown on Figure 11.29.

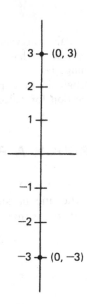

Figure 11.29

In general, let $A = (a_1, a_2)$ be an arbitrary point. We define $-A$ to be the points $(-a_1, -a_2)$.

Example. Let $A = (1, 2)$. Then $-A = (-1, -2)$. We represent $-A$ in Figure 11.30.

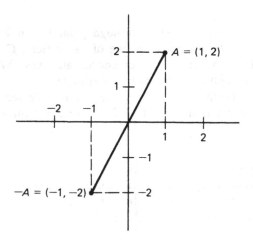

Figure 11.30

Example. Let $A = (-2, 3)$. Then $-A = (2, -3)$. We draw A and $-A$ in Figure 11.31.

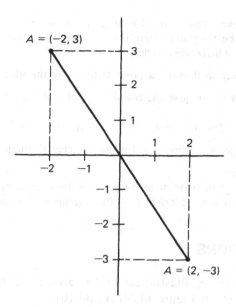

Figure 11.31

In each case we see that $-A$ is the reflection of the point A through the origin O. Thus in terms of coordinates, if R denotes reflection through O, then we have

$$R(A) = -A.$$

11, §5. EXERCISES

1. Copy the points below. Draw the images of points S, T, and P when reflected through the point O.

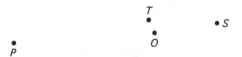

Figure 11.32

2. Copy rectangle $ABCD$ and point O onto your paper.
   (a) Draw the image of the rectangle when reflected through O.

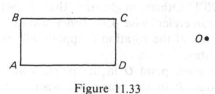

Figure 11.33

(b) Draw its image when reflected through point A.

(c) Let point X be the point where the diagonals of $ABCD$ intersect. What is the image of $ABCD$ when reflected through X?

3. Redefine the reflection through a point O by using the idea of midpoint.

4. For each point A below, give the coordinates of $-A$, its reflection through the origin:

   (a) $A = (7, 56)$ (b) $A = (-4, -7)$ (c) $A = (3, -8)$

5. What is the image of the line $x = 4$ when reflected through the origin?

6. **Exercise in 3-space**

   Let $P = (p_1, p_2, p_3)$ be a point in 3-space. Write a suitable definition for the reflection of P through the origin $(0, 0, 0)$ in terms of the coordinates of P.

11, §6. ROTATIONS

We shall start with an illustration before giving the formal definition. Consider the objects in Figure 11.34(a) and (b):

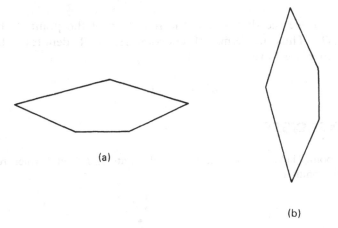

(a)

(b)

Figure 11.34

Do these objects have the same shape? Most people would say "yes", but that the object in Figure 11.34(b) is "rotated counterclockwise a quarter turn, or 90°". Some people might say 11.34(b) is 11.34(a) rotated *clockwise* 270°! Others might say that 11.34(b) is the same as 11.34(a) rotated counterclockwise 450° (once around 360°, and then 90° more)! Our definition of the rotation mapping will eventually encompass all of these possibilities.

We start with a given point O in the plane, which will be a reference point. For any point P in the plane, we wish to find its image when rotated around O by some amount. Roughly speaking, rotations work like a record player: point O is like the center spindle, while the rest of the plane rotates around it like a record. We can determine the amount of rotation by giving an angle measure between $0°$ and $360°$, and specifying whether we move in the clockwise or counterclockwise direction.

The rotation mapping can then be defined as follows. Given a point O and a number x such that $0 \leqq x < 360$. For any point P in the plane, draw a circle with center O and radius $|OP|$ (Figure 11.35). Starting at P, move along the circle *counterclockwise* until you locate the first point P' such that

$$m(\angle POP') = x°.$$

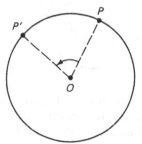

Figure 11.35

The association $P \mapsto P'$ is the **rotation by $x°$ with respect to O**. Since we have already used the letter R for reflection, we choose another letter, G, for rotation, and we denote the rotation by $x°$ with respect to a point O as: $G_{x,O}$. Figure 11.36 shows the rotation of a flower $90°$ with respect to O.

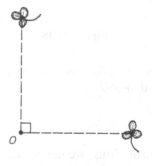

Figure 11.36

If the point O is fixed throughout a discussion, and all rotations are with respect to O, then we often omit O from the notation and write G_x instead of $G_{x,O}$.

So far, we have defined rotation only in a counterclockwise direction. We can also define rotation in a clockwise direction. Given a point O and a number x such that $0 \leqq x < 360$. For any point P, we find the point P' by drawing a circle through P with center O, and moving along the circle *clockwise* until we hit the first point where $m(\angle POP') = x°$, as illustrated:

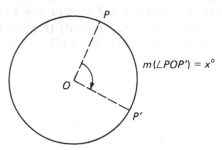

Figure 11.37

We denote this clockwise rotation by $G_{-x, O}$. Notice the minus sign! A minus sign in front of the angle measure means rotate in the *clockwise* direction. Figure 11.38 illustrates a flower rotated by $-45°$ with respect to O.

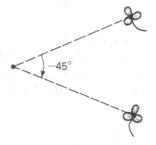

Figure 11.38

This convention allows us to talk about a rotation by any number of degrees between $-360°$ and $+360°$.

Equality of Mappings

Before considering more rotations, we define what it means for two mappings to be equal:

A mapping F and a mapping G are **equal** if and only if

$$F(P) = G(P)$$

for *every* point P in the plane.

Example. Let O be a point in the plane. Let F be reflection through point O, and let G be rotation by 180° around O. Draw a point O on your paper, pick out some arbitrary locations for point P, and convince yourself that $F(P)$ is the same point as $G(P)$ in every case.

When two mappings F and G are equal, as in the above case, we write: $F = G$.

Notice that the definition of equality depends on the *value* of the mappings at each point, *not* on the descriptions of the mappings. In the example above, we're reflecting points in one case, and rotating them in another; but the effect is the same for all points, and so the mappings are equal.

Consider a rotation by 0°. In this case, the image of any point P in the plane is just P itself. This is a special mapping, called the **identity mapping**. By definition, the identity mapping associates to each point P the point P itself. This mapping is denoted by the letter I. Thus we have:

$$I(P) = P$$

for every point P in the plane.

According to our definition of equality given above, we have:

$$I = G_{0°}.$$

Finally, we can even define what we mean by rotations by *any* number, not just those between -360 and 360. For example, consider $G_{500°}$, a rotation by 500°(!). We can interpret this as a rotation by 360° (once around back to start) with an additional rotation by 140°. See Figure 11.39.

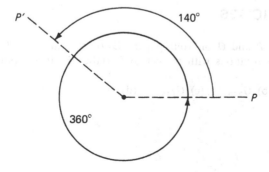

Figure 11.39

Thus we see that

$$G_{500°} = G_{140°}.$$

In a similar manner, $G_{860°}$ is a rotation twice around (720° worth) plus an additional 140°. Thus we have

$$G_{860°} = G_{140°} = G_{500°}.$$

The same kind of interpretation works for negative number rotations, except that these go in the clockwise direction. For example,

$$G_{-450°} = G_{-90°} = G_{270°}.$$

Notice that if we rotate the plane by 360° then every point P comes back to itself. Thus according to our convention about equality of mappings, we have the equality

$$G_{0°} = G_{360°} = I.$$

Sometimes people point out that rotating by 360° is not "the same" as rotating by 0°, because rotation by 360° involves a "motion". Nevertheless, it has been found convenient to use the terminology about equality of mappings as we have done.

It is too involved for this course to give a description of rotations in terms of coordinates in general, and so we shall omit this. It is however easy in special cases, like rotation by 180° or 360°, and we leave this as an exercise (cf. Exercises 10–12).

11, §6. EXERCISES

1. Copy points P and O on your paper. Draw the image of P under each of the following rotations with respect to O (label the image points P_a, P_b, P_c, etc.):
   (a) G_{90} (b) G_{270} (c) G_{135} (d) G_0

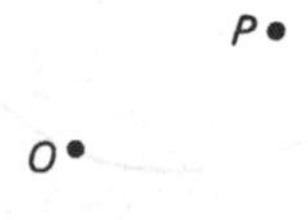

$$P \bullet$$

$$O \bullet$$

Figure 11.40

2. Give an equivalent rotation between 0 and 360 for each of the following rotations:

    (a) G_{400} (b) G_{-75} (c) G_{1080} (d) G_{-500} (e) G_{-780}

3. Draw a point O and a point P on your paper. Draw the image of P under each of the rotations given in Exercise 2 with respect to O.

4. Draw an equilateral triangle XYZ on your paper. Draw its image when rotated by $120°$ with respect to point X.

5. Draw the point O and the "L" shape on your paper. Draw the image of the L when rotated by $270°$ with respect to O.

O

Figure 11.41

6. Show that rotation by $180°$ or $-180°$ with respect to a point O is equal to reflection through point O. (Choose an arbitrary point X, and show that the image of X is the same whether you rotate or reflect.)

7. Which rotations are equal to the identity map?

8. Let x and y be numbers, and suppose that rotation $G_{x,O} = G_{y,O}$. What can you say about the numbers x and y. (Obviously, they may be equal, but the rotations can be equal without x and y being equal. What then must be true about x and y?)

9. Let x be a number greater than $360°$. Devise a method that will give a number y which is between 0 and 360 ($0 \leq y < 360$) such that $G_x = G_y$. Write your method out precisely as if you were writing it up for a book.

10. Let $P = (4, 0)$ be a point in the plane. Write the coordinates of the image of P under each of the rotations given below. All the rotations are with respect to the origin.

    (a) $G_{90°}$ (b) $G_{180°}$ (c) $G_{-270°}$ (d) $G_{360°}$

11. Repeat Exercise 10 with $P = (0, -6)$.

12. Repeat Exercise 10 above with $P = (3, 6)$. Use a piece of graph paper to get a clear idea of what is going on.

13. A triangle has vertices $P = (3, 2)$, $Q = (3, -2)$, and $R = (6, 0)$.
    (a) Write down the coordinates of the images P', Q', and R' when the triangle is reflected through the y-axis.
    (b) Can triangle PQR be mapped onto triangle $P'Q'R'$ by a rotation? If yes, what rotation?

14.

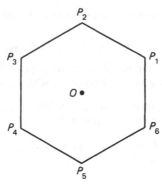

Figure 11.42

(a) How many lines of symmetry does the regular hexagon have?
(b) Consider a rotation of the regular hexagon around its center point O. By how many degrees would you have to rotate it in order to map point P_1 onto point P_3?
(c) Now consider a rotation about the center point of an arbitrary *regular n-gon*. If the n-gon is labeled in the same way (counterclockwise, P_1 to P_n), how many degrees would you have to rotate it around its center in order to map point P_1 onto point P_3? Your answer should have an n in it.
(d) Back to the regular hexagon. Now consider a rotation around point P_2. How many degrees would you have to rotate it in order to map point P_1 onto point P_3?

11, §7. TRANSLATIONS

A translation "moves" a point a particular distance in a given direction.

The easiest way to specify a direction and distance is to designate two points in the plane as reference points, and draw an arrow between them, as in Figure 11.43.

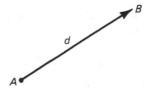

Figure 11.43

This arrow gives a direction, and its length gives us a distance. The arrow drawn in Figure 11.43 starts at A, and B is its end point. A pair of points, where A is the beginning point and B is the end point, is called a **vector,** and is denoted by $\overrightarrow{AB}$. If we draw the vector starting at B and pointing to A, as shown in Figure 11.44, we would denote it $\overrightarrow{BA}$.

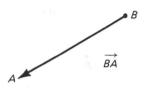

Figure 11.44

We now can define a translation mapping. Given a vector $\overrightarrow{AB}$ with length d. To each point P in the plane, we associate the point P' which is at distance d from P in the direction the vector points. This association is called **translation** by $\overrightarrow{AB}$, which we denote by T_{AB}. We illustrate a point P and point $P' = T_{AB}(P)$:

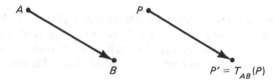

Figure 11.45

Note that for any point P, the vector $\overrightarrow{PP'}$ is parallel to and has the same length as the reference vector $\overrightarrow{AB}$.

If we are willing to use Theorem 7-5, it follows that the points A, P, P', B are the four vertices of a parallelogram, because the two opposite sides $\overrightarrow{AB}$ and $\overrightarrow{PP'}$ of the quadrilateral $APP'B$ are parallel and have the same length.

Figure 11.46 illustrates the translation T_{BA}.

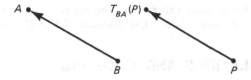

Figure 11.46

Notice that $T_{AB}(A) = B$ and that $T_{BA}(B) = A$. What mapping do you think is equal to T_{AA}?

11, §7. EXERCISES

1. Copy points A, B, P, and Q onto your paper.

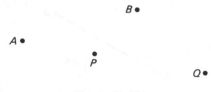

Figure 11.47

Draw and label each of the following points:
(a) $T_{AB}(P)$ (b) $T_{AB}(Q)$ (c) $T_{AB}(A)$ (d) $T_{QP}(B)$ (e) $T_{QP}(P)$

2. Draw a triangle $\triangle PQR$ on your paper. Draw the image of $\triangle PQR$ under the translation T_{PQ}.

3. Draw three points on your paper (not on the same line), and label them A, B, and X. What figure is formed by joining A, B, $T_{AB}(X)$, and X?

4. The border design illustrated below has the property that it can be "translated along itself" and still "look the same". What we mean is that the image of the design under the translation T_{AB} (for example) will coincide with the original figure. Draw two other border designs which have this same property.

Figure 11.48

5. Let T be a translation. Prove from Euclid's tests that for two points P, Q we have

$$d(P, Q) = d(T(P), T(Q)).$$

In other words, the distance between two points is equal to the distance between the translated points. [*Hint*: Use Theorem 7-5 of Chapter 7, §2.]

11, §8. TRANSLATIONS AND COORDINATES

We shall be able to give a coordinate definition of T_{AB} where A and B are any points in the plane in a moment. The next Experiment illustrates an important property that we will use.

EXPERIMENT 11-3

1. Let $A = (2, 4)$ and $B = (5, 8)$. On a piece of graph paper, draw A, B and $\overrightarrow{AB}$. Compute $(B - A)$, and locate it on the graph paper. Draw $\overrightarrow{O(B - A)}$.

2. Repeat Part 1 using $A = (-2, 3)$ and $B = (4, -4)$.

3. In each case above, the points O, A, B, and $(B - A)$ lie on the corners of what geometric figure?

4. How do the lengths of vectors $\overrightarrow{AB}$ and $\overrightarrow{O(B - A)}$ compare?

5. How do the directions of vectors $\overrightarrow{AB}$ and $\overrightarrow{O(B - A)}$ compare?

6. Consider the translation $T_{O(B-A)}$. How does this mapping compare with translation T_{AB}?

In the light of the Experiment, we see that translations described in §7 and the addition of points are related. This allows us to define translations using coordinates.

Let A be a point in $\mathbf{R}^2$. We first define translation by vector $\overrightarrow{OA}$. We see that translation by $\overrightarrow{OA}$ is the association which to each point P of the plane associates the point $P + A$. In Figure 11.49, we have drawn the effect of this translation on several points.

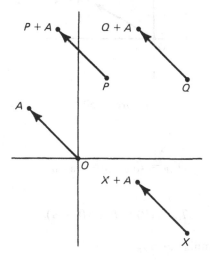

Figure 11.49

The association

$$P \mapsto P + A$$

has been represented by arrows in Figure 11.49.

Instead of writing T_{OA} we shall write more simply T_A to denote translation by $\overrightarrow{OA}$. We shall also call it **translation by** A. Thus the value of T_A at a point P is

$$T_A(P) = P + A.$$

Example. Let $A = (-2, 3)$ and $P = (1, 5)$. Then

$$T_A(P) = P + A = (-1, 8).$$

To describe T_{AB} for general A, B we observe that the vectors

$$\overrightarrow{AB} \quad \text{and} \quad \overrightarrow{O(B - A)}$$

are parallel, have the same direction and same length, as illustrated on Figure 11.50, and in Experiment 11-3.

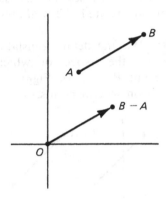

Figure 11.50

Thus

$$T_{AB} = T_{O(B-A)} = T_{(B-A)}.$$

Since

$$T_{(B-A)}(P) = P + (B - A),$$

we can write the formula for T_{AB}.

Theorem 11-1. *If $A, B,$ and P are points in $\mathbf{R}^2$, then*

$$T_{AB}(P) = P + B - A.$$

Example. Let $A = (1, 2)$, $B = (2, 3)$, and $P = (-1, 3)$. Then

$$B - A = (1, 1)$$

and

$$T_{AB}(P) = (-1, 3) + (1, 1) = (0, 4).$$

This is illustrated on Figure 11.51.

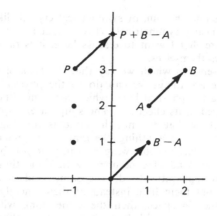

Figure 11.51

Theorem 11-1 gives a simple test to determine when two translations are equal.

Theorem 11-2. *Translations T_{AB} and T_{CD} are equal if and only if*

$$B - A = D - C.$$

Proof. Suppose $T_{AB} = T_{CD}$. Then for every point P, we have by Theorem 11-1:

$$P + B - A = P + D - C.$$

Subtracting P from both sides yields $B - A = D - C$. Conversely, if $B - A = D - C$, then Theorem 11-1 shows that $T_{AB}(P) = T_{CD}(P)$ for all points P, so $T_{AB} = T_{CD}$.

Example. Let $A = (3, 1)$, $B = (-1, 2)$, $C = (4, 5)$, and $D = (1, -3)$. Determine whether $T_{AB} = T_{CD}$.
We have

$$B - A = (-4, 1) \qquad \text{and} \qquad D - C = (-3, -8).$$

Since $B - A \neq D - C$ we conclude that $T_{AB} \neq T_{CD}$.

To see how some of the notions discussed in this chapter apply to the physical world, we suggest that you read Chapter 4 of Richard Feynman's book *The Character of Physical Law*. In that chapter Feynman discusses "Symmetry in physical law", and we could not write up such a discussion better than he does. We shall quote a few lines from his chapter just to give you the flavor:

> Symmetry seems to be absolutely fascinating to the human mind. We like to look at symmetrical things in nature, such as perfectly symmetrical

spheres like planets and the sun, or symmetrical crystals like snowflakes, or flowers which are nearly symmetrical. However, it is not the symmetry of the objects in nature that I want to discuss here; it is rather the symmetry of the physical laws themselves....

... That is the sense in which we say that the laws of physics are symmetrical; that there are things we can do to the physical laws, or to our way of representing the physical laws, which make no difference, and leave everything unchanged in its effects.... The simplest example of this kind of symmetry—you will see that it is not the same as you might have thought, left and right symmetric, or anything like that—is a symmetry called translation in space. This has the following meaning: if you build any kind of apparatus, or do any kind of experiment with some things, and then go and build the same apparatus to do the same kind of experiment, with similar things but put them here instead of there, merely translated from one place to another in space, then the same thing will happen in the translated experiment as would have happened in the original experiment....

Let us take as an example the law of gravitation, which says that the force between objects varies inversely as the square of the distance between them; ... If I have a pair of objects like a planet going around a sun, and I move the whole pair over, then the distance between the objects of course does not change, and so the forces do not change. Further, when they are in the moved-over situation they will go at the same speed, and all the changes will remain in proportion and everything go around in the two systems in exactly the same way. The fact that the law says "the distance between two objects", rather than some absolute distance from the central eye of the universe, means that the laws are translatable in space.

That, then, is the first symmetry translation in space. The next one could be called translation in time, but, better, let us say that delay in time makes no difference....

Let us take some other examples of symmetry laws. One is a rotation in space, a fixed rotation. If I do some experiments with a piece of equipment built in one place, and then take another one (possibly translated so that it does not get in the way) exactly the same, but turned so that all the axes are in a different direction, it will work the same way. Again we have to turn everything that is relevant. If the thing is a grandfather clock, and you turn it horizontal, then the pendulum will just sit up against the wall of the cabinet and not work. But if you turn the earth too (which is happening all the time) the clock still keeps working.

The mathematical description of this possibility of turning is a rather interesting one....

Now go read Feynman.

11, §8. EXERCISES

1. Let $A = (-2, 5)$. Find $T_A(P)$ for each point P given below:
   (a) $P = (1, 5)$ (b) $P = (-3, -6)$ (c) $P = (2, -5)$
   (d) $P = (0, 0)$ (e) $P = (p_1, p_2)$

In the next four exercises, we consider the translation T_A where $A = (-3, 2)$.

2. Let S be the triangle with vertices $(2, 5)$, $(-3, 7)$, and $(3, 6)$. What are the coordinates of the vertices of triangle $T_A(S)$, the image of S under translation T_A?

3. Let L be the line $x = -4$. Describe the image of L under the translation T_A.

4. Let K be the line $y = 4$. Describe the image of K under the translation T_A.

5. Let C be a circle centered at the origin with radius 3. What are the coordinates of the center of the circle $T_A(C)$? What is the radius of $T_A(C)$?

6. Let $A = (0, 0)$ $B = (2, 3)$ $C = (4, 6)$
         $D = (-2, -3)$ $E = (3, 2)$ $F = (5, 5)$
State whether the following statements are true or false:
(a) $T_{AB} = T_{BA}$ (e) $T_{AB} = T_{EC}$
(b) $T_{AB} = T_{BC}$ (f) $T_{BC} = T_{EF}$
(c) $T_{BC} = T_{AD}$ (g) $T_{EC} = T_{EF}$
(d) $T_{BC} = T_{DA}$ (h) $T_{DB} = T_{AC}$

ADDITIONAL EXERCISES FOR CHAPTER 11

1. Let F be a mapping of the plane into itself. We define a **fixed point** of F to be a point P such that $F(P) = P$. For example, let F be rotation by 90° around a given point O. Then O is a fixed point since $F(O) = O$. State whether each of the following mappings have any fixed points. If yes, say how many and where they are:
(a) The identity mapping.
(b) Reflection through a given point O.
(c) Reflection through a given line L.
(d) A rotation not equal to the identity, with respect to a given point O.
(e) A translation.
(f) The constant mapping whose value is a given point X.

2. Two houses A and B and a power line L are situated as indicated below. What is the minimum length of wire required to bring power to both houses if there is to be only one transformer at the power line and separate wires running to each house from the transformer.

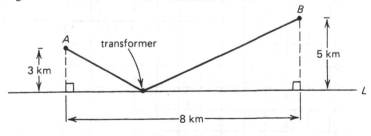

Figure 11.52

[*Hint*: Reflect A through line L; connect A' and B.]

3. There are an infinite number of mappings; we have just looked at some of them. It's easy to make up others. For example, draw a line L on the plane.

For any point P on the plane, let P' be the point where a line through P perpendicular to L intersects L (see illustration):

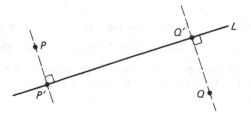

Figure 11.53

The association $P \mapsto P'$ defines a mapping, called the **perpendicular projection on the line** L.

Make up three mappings of your own. Keep in mind this important property of mappings: two or more points may have the same image point (like the constant map, or the one given just above) but each point may have only *one* image. There is no room for ambiguity!! When you make up a mapping, be sure that you describe exactly where the image of each point is located.

For each of your three mappings, draw the image of some points and of a triangle. Also, tell where the fixed points are, if there are any.

The next six questions refer to the graph below.

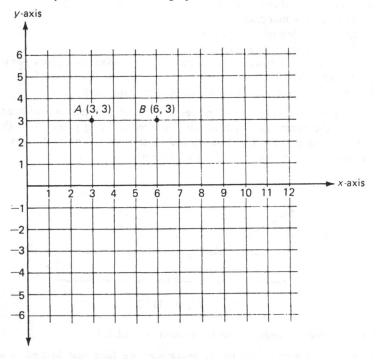

Figure 11.54

4. Reflect point A through the x-axis. Label the image A' and write down its coordinates.

5. Translate point B with respect to vector $\overrightarrow{AB}$. Label the image point B' and write down its coordinates.

6. Rotate point B' around point B by $90°$. Label the image B'' and write down its coordinates.

7. Find the value of $d(A', B'')$.

8. What is the area of $\triangle AB''B'$?

9. What is the measure of $\angle AB''B'$?

10. Given points A and B, between lines L_1 and L_2. Illustrate and carefully describe how you would find the *shortest* path from A to L_1 to L_2 to B.

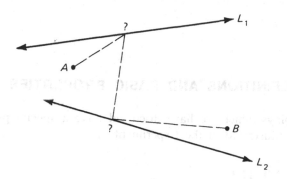

Figure 11.55

[*Hint*: Reflect A through line L_1 and reflect B through line L_2.]

CHAPTER 12

Isometries

12, §1. DEFINITIONS AND BASIC PROPERTIES

Some mappings which we have discussed have a special property which will first be illustrated in the Experiment.

EXPERIMENT 12-1

1. On your paper, draw three points X, Y, and Z which do not lie on a straight line. Draw segments $\overline{XY}$ and $\overline{YZ}$, so that your picture looks something like:

Figure 12.1

2. Using a ruler, find $d(X, Y)$ and $d(Y, Z)$. Write these down. Record also the measure of $\angle XYZ$.

3. Draw a line L on your paper (it can be anywhere). Now reflect points X, Y, and Z through L, using ruler and compass as given in Construction 11-1. Mark the images of these points X', Y', and Z' respectively.

4. Measure and write down $d(X', Y')$ and $d(Y', Z')$. Draw segments $\overline{X'Y'}$ and $\overline{Y'Z'}$ and measure angle $\angle X'Y'Z'$.

5. How do the measurements in Part 4 compare with those in Part 2?

On a new piece of paper draw and label four points X, Y, Z, and O. Repeat Parts 1 through 5, except this time rotate the points X, Y, and Z around point O by $x°$ (you pick a value for x). Use compass and protractor to get accurate diagrams.

On a new piece of paper, draw a vector $\overrightarrow{AB}$ and points X, Y, and Z. Repeat Parts 1 through 5, using the translation determined by the vector rather than a rotation or reflection. Use a ruler to find the images of these points as accurately as possible.

Repeat Parts 1 through 5 one more time, except this time dilate the points X, Y, and Z by 3 with respect to a point O. Again label the images X', Y', and Z', and measure distances as well as the original and dilated angles.

What conclusions can you reach about these mappings?

6. Draw two parallel lines, L and K. Pick a few points on L and a few on K. Draw a third line M, and reflect the points you've chosen through line M (use ruler and compass). Where are the images of the rest of the points on lines L and K?

7. Repeat Part 6 except rotate the points on lines L and K around a point O (you choose a number of degrees).

8. If two lines are parallel, and we rotate or reflect them, what can we say happens?

We can now define the special property of these mappings.

Let F be a mapping, and suppose P and Q are two points in the plane. When is the distance between $F(P)$ and $F(Q)$ going to be the same as the original distance between P and Q? In other words, when does

$$d(P, Q) = d(F(P), F(Q)) \quad ??$$

The Experiment should have given you a clue. When F is one of the following mappings, then these distances WILL be the same:

reflection through a line,

rotation,

translation.

We say that a mapping F **preserves distances** or is **distance preserving** if and only if:

for every pair of points P, Q in the plane, the distance between P and Q is the same as the distance between $F(P)$ and $F(Q)$.

Such a mapping is called an **isometry** ("iso" is a prefix meaning "same", and "metry" means measure). In other words, we define an **isometry** to be a mapping F such that for every pair of points P, Q in the plane we have

$$d(P, Q) = d(F(P), F(Q)).$$

Roughly speaking, isometries are mappings which do not "distort" figures in the plane. The distance between points is not disturbed. Sometimes isometries are referred to as "rigid" mappings.

We shall accept without proofs (so we accept as postulate) the following property:

The mappings reflection through a line, rotation, and translation, are isometries.

Remark. Actually, Exercise 29 of Chapter 3, §2 proved the statement for reflections using only Postulate **RT**. Using Euclid's **SAS** postulate, you can prove that a rotation is an isometry. Conversely, on the other hand, it is possible to give foundations of the theory of isometries by avoiding completely the development we have given in most of this book, and to prove directly that rotations, translations, and reflections are isometries. Then one can prove Euclid's tests for congruence, as will be shown in §5.

For instance, in Exercise 4 of §2, using the definition of translations by means of coordinates, you will prove directly that a translation is an isometry.

For reflection through a line, by a suitable choice of coordinate system, one can assume that the line is the x-axis, for instance. Then reflection is just the mapping

$$(x, y) \mapsto (x, -y).$$

In this case, you can prove directly from the definition of distance in Exercise 5 of §2 that the reflection is an isometry.

In the same way, you can prove that reflection through a point is an isometry. By a suitable choice of coordinate system, you can assume that the point is the origin $(0, 0)$, and then reflection through the origin is simply the mapping

$$A \mapsto -A.$$

From the definition of distance in terms of coordinates, you can prove directly in Exercise 6 of §2 that

$$d(A, B) = d(-A, -B).$$

Similarly, one can give a definition of rotations using only numbers, and one could verify that a rotation is an isometry. The algebra needed to do so is more complicated, and we won't do it in this course.

We also accept without proof:

Isometries preserve the measure of an angle.

Remark. *Let F be an isometry. If P and Q are distinct points, then F(P) and F(Q) must be distinct.*

We can prove this easily. The distance between P and Q is not 0, therefore the distance between $F(P)$ and $F(Q)$ cannot be 0 either (remember isometries preserve distances). Thus $F(P) \neq F(Q)$. (Recall **DIST 1** in Chapter 1.)

There are two other important properties of isometries which will be taken as postulates. In the exercises, you will check experimentally that they are true for reflections, rotations, and translations by using constructions.

ISOM 1. *Let F be an isometry. The image of a line segment under F is a line segment. In other words, if we take the image under F of each point on a line segment (by rotating, reflecting, or whatever), we get a set of image points which themselves make up a line segment.*

Exercise. If you are theoretically inclined, use the postulate **SEG** from Chapter 1, §2 to prove that if F is an isometry, and M is a point on the line segment $\overline{PQ}$, then $F(M)$ is a point on the line segment between $F(P)$ and $F(Q)$.

Example. Given line segment $\overline{PQ}$, if we reflect it through line L, we get another line segment $P'Q'$ as shown on the figure.

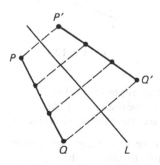

Figure 12.2

ISOM 2. *Let F be an isometry. The image of a line under F is a line.*

12, §1. EXERCISES

1. On your paper, draw a line segment $\overline{PQ}$ and a line L. Choose four or five points on $\overline{PQ}$, and carefully reflect them through line L. Observe that the image points also lie along a line segment.

2. Repeat Exercise 1, except rotate the points on $\overline{PQ}$ around a point O.

3. Repeat Exercise 1, except translate the points on $\overline{PQ}$ by some vector.

4. Draw the image of a circle of radius r, center P under
   (a) reflection through its center;
   (b) reflection through a line L outside of the circle, as in Figure 12.3(a);
   (c) rotation by $90°$ with respect to a point O outside the circle as in Figure 12.3(b);
   (d) rotation by $270°$ with respect to O;
   (e) translation.

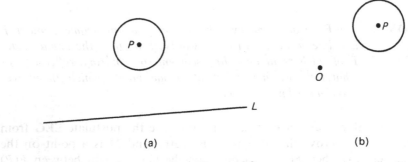

(a)   (b)

Figure 12.3

5. Given points Q and M as shown:

Figure 12.4

Let F be an isometry, and suppose $F(Q) = Q$.
(a) Draw all the possible locations for $F(M)$.
(b) Explain why these are the only possible locations.

   In Exercise 5, you have to explain two things. First, why the points on your drawing are possible locations. Show that for each point of your drawing, there is an isometry F such that $F(Q) = Q$ and such that $F(M)$ is equal to that point. Second, you have to explain why there are *no other* possible locations.

6. Given points P, Q, and M as shown, with M on the segment $\overline{PQ}$.

Figure 12.5

Let F be an isometry. Suppose $F(P) = P$ and $F(Q) = Q$.
(a) Draw the possible locations for $F(M)$.
(b) Explain why these are the only possible locations. [Here, you may find it useful to use Postulate **SEG** from Chapter 1, §2.]

7. Given points P, Q, M as shown, with M on the line through P, Q,

Figure 12.6

Let F be an isometry. Suppose $F(P) = P$ and $F(Q) = Q$.
(a) Draw the possible locations for $F(M)$.
(b) Explain why these are the only possible locations.

8. Let L and K be two parallel lines, and let F be an isometry. Prove that $F(L)$ and $F(K)$ are parallel. [*Hint*: Assume that $F(L)$ and $F(K)$ are *not* parallel; then they intersect. Use the definition of isometry to deduce a contradiction.]

9. Let L and K be two perpendicular lines, and let F be an isometry. Prove that $F(K)$ and $F(L)$ are perpendicular.

10. Using the result from Exercise 9, prove that if F is an isometry and R is a rectangle, then $F(R)$ has the same area as R.

11. Suppose that the two segments $\overline{PQ}$ and $\overline{MN}$ meet in a point O and bisect each other. Prove that $d(P, M) = d(Q, N)$, by finding an isometry between $\overline{PM}$ and $\overline{QN}$.

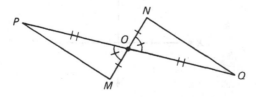

Figure 12.7

12. In Figure 12.8, we suppose that

$$|WO| = |YO|, \qquad |VO| = |ZO|$$

and that $m(\angle VOW) = m(\angle ZOY)$. Prove that

$$|ZY| = |VW|$$

by using an isometry which maps Z on V and Y on W.

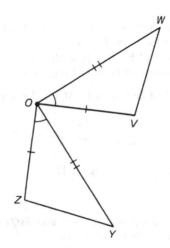

Figure 12.8

13. Line L is the perpendicular bisector of $\overline{PQ}$ and $\overline{XY}$. Prove that

$$d(P, X) = d(Q, Y)$$

by using an isometry which maps P on Q and X on Y.

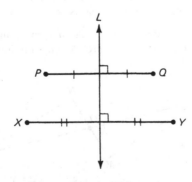

Figure 12.9

14. Let P_y be the mapping which maps a point (x, y) in the coordinate plane onto the point $(0, y)$. This mapping is called the **projection onto the y-axis**. For example, if $X = (4, -2)$, we have

$$P_y(X) = (0, -2).$$

See Figure 12.10 below.

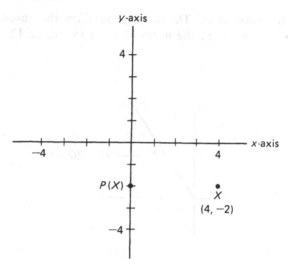

Figure 12.10

(a) Let $A = (-8, 9)$ and let $B = (1, 3)$. Write the coordinates of $P_y(A)$ and $P_y(B)$.
(b) Is the mapping P_y an isometry?
(c) Let $X = (x, y)$ and $V = (u, v)$ be arbitrary points. Let $X' = P_y(X)$ and let $V' = P_y(V)$. Write a formula which gives $d(X', V')$.
(d) Suppose A and B are any two points in the plane, and that $P_y(A) = P_y(B)$. What can you conclude about A and B?
(e) Suppose T is a triangle in the plane. What is the image of T under the mapping P_y?

12, §2. RELATIONS WITH COORDINATES

Since we have introduced addition and subtraction for points, we shall now describe a way of expressing the distance between points by using subtraction.

We recall that the distance between two points P, Q is denoted by $d(P, Q)$. We shall use a special symbol for the distance between a point and the origin, namely the absolute value sign. We let

$$d(A, O) = |A|.$$

Thus we use two vertical bars on the sides of A. If $A = (a_1, a_2)$, then

$$|A| = \sqrt{a_1^2 + a_2^2},$$

and therefore

$$|A|^2 = a_1^2 + a_2^2.$$

We call $|A|$ the **norm** of A. The norm generalizes the absolute value of a number. We can represent the norm of A as in Figure 12.11.

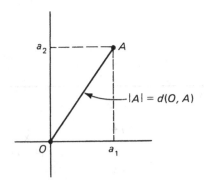

Figure 12.11

Note that $|A| = |-A|$. See Exercise 3.

Using our subtraction of points, we can express the distance between two points A, B by

Theorem 12-1.

$$d(A, B) = |A - B| = |B - A|.$$

It is easy to see that this is true. Let $A = (a_1, a_2)$ and let $B = (b_1, b_2)$. Then

$$B - A = (b_1 - a_1, b_2 - a_2)$$

and

$$|B - A| = \sqrt{(b_1 - a_1)^2 + (b_2 - a_2)^2}.$$

But the right-hand side of the equation above is precisely $d(A, B)$ according to our distance formula. Thus we have

$$|A - B| = d(A, B).$$

In a similar manner, we may prove that

$$|B - A| = d(A, B) \quad \text{as well.}$$

Note that this tells us that the length of an arbitrary vector $\overrightarrow{AB}$ is equal to the length of vector $\overrightarrow{O(B - A)}$ or $\overrightarrow{O(A - B)}$:

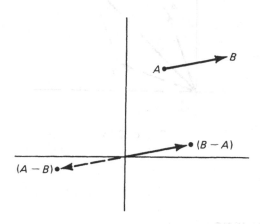

Figure 12.12

a fact which we recognized earlier, in the chapter on translations.

This "norm" notation makes our work with isometries much neater. For example, we may write the definition of isometry as follows:

A mapping F of the plane into itself is an isometry if and only if for every pair of points P, Q we have

$$|F(P) - F(Q)| = |P - Q|.$$

Many distance calculations become simplified when distances are expressed in norm notation. For example,

$$d(A, A + B) = |(A + B) - A|$$

$$= |(B + A) - A| \quad \text{since addition of points is commutative}$$

$$= |B + (A - A)| \quad \text{since addition of points is associative}$$

$$= |B|.$$

This demonstration confirms our intuition when using the parallelogram law:

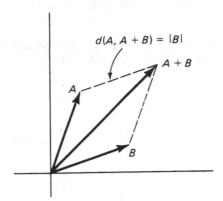

Figure 12.13

12, §2. EXERCISES

1. For each of the points P below, find $|P|$:
   (a) $P = (-1, 3)$ (b) $P = (137, 137)$ (c) $P = (1, -3)$

2. For each pair of points A, B given below, find $|A - B|$.
   (a) $A = (3, 6)$, $B = (3, -4)$ (b) $A = (-2, 4)$, $B = (-3, -5)$
   (c) $A = (0.5, 7)$, $B = (-1.5, 4)$ (d) $A = (\sqrt{2}/2, \sqrt{2}/2)$, $B = (0, 0)$

3. Let $X = (x_1, x_2)$ be any point. Using the distance formula, prove that

$$|X| = |-X|.$$

4. Let T_A be the translation by point A. Prove that T_A is an isometry using the following steps:
   (a) Let P and Q be arbitrary points. Express the distance between P and Q using norm notation.
   (b) What is $T_A(P)$? $T_A(Q)$?
   (c) Express the distance between $T_A(P)$ and $T_A(Q)$ using norm notation, and show that it equals the distance between P and Q.

5. Prove that reflection through the x-axis is an isometry. (Take two arbitrary points, compute the distance between them, etc.)

6. Prove that reflection through O preserves distances. In other words, prove that

$$d(A, B) = d(-A, -B).$$

12, §3. COMPOSITION OF ISOMETRIES

It is possible to create new distance-preserving mappings (isometries) out of the ones we already know. Experiment 12-2 suggests how to do this.

EXPERIMENT 12-2

Part I

Draw a line on your paper and label it L. Pick a point O not on L. Pick another point X at random.

1. Let R_L be reflection through line L. Find $R_L(X)$. Label the image of X under R_L by X'.

2. Let G be the rotation around point O by $90°$. Find the point $G(X')$ and label it X''. In other words, rotate X' (not X) around O by $90°$. The association

$$X \mapsto X''$$

   is a new mapping, which is a combination of a reflection, followed by a rotation. Is this an isometry? Test it out.

3. Pick another random point Y. Let $Y' = R_L(Y)$ be its reflection through line L. Let Y'' be the rotation of Y' around O by $90°$. Measure the distances

$$d(X, Y) \quad \text{and} \quad d(X'', Y'').$$

   Are they equal? Perform Parts 1, 2, 3 several times for at least four choices of X and Y.

Part II

Use the same points and lines as before.

4. Let G be rotation by $90°$ around O again. Find $G(X)$ and label it $\bar{X}$.

5. Reflect $\bar{X}$ through L and label this reflection $\bar{\bar{X}}$.
   The association

$$X \mapsto \bar{\bar{X}}$$

   is a mapping. Is $\bar{\bar{X}} = X''$? Is this new mapping equal to the one found in Part 2?

6. Using the same point Y as before, rotate Y around O by 90° to obtain the point $\bar{Y}$. Then reflect $\bar{Y}$ through L and label this reflection $\bar{\bar{Y}}$. Measure the distances

$$d(X, Y) \quad \text{and} \quad d(\bar{\bar{X}}, \bar{\bar{Y}}).$$

Are those distances equal? Is the mapping

$$X \mapsto \bar{\bar{X}}$$

an isometry?

Part III

Take a new piece of paper, and choose four random points X, Y, Z, and O. Let G_{40} be rotation around O by 40°, and let G_{60} be rotation around O by 60°.

7. Find $G_{40}(X)$, $G_{40}(Y)$, and $G_{40}(Z)$ and label these points X', Y', and Z'.

8. Find $G_{60}(X')$, $G_{60}(Y')$, and $G_{60}(Z')$ and label these points X'', Y'', and Z''.

Compare the locations of X'', Y'', and Z'' with points X, Y, and Z. Again we have created a new mapping $X \mapsto X''$ by combining two rotations.

Part IV

For the following questions, use the same points as in Part III.

9. Find $G_{60}(X)$, $G_{60}(Y)$, and $G_{60}(Z)$. Label the points $\bar{X}$, $\bar{Y}$, and $\bar{Z}$.

10. Find $G_{40}(\bar{X})$, $G_{40}(\bar{Y})$, and $G_{40}(\bar{Z})$, and label the points $\bar{\bar{X}}$, $\bar{\bar{Y}}$, $\bar{\bar{Z}}$.

Again we have created a new mapping $X \mapsto \bar{\bar{X}}$ by combining the rotations, but in the reverse order.

11. Does $X'' = \bar{\bar{X}}$, $Y'' = \bar{\bar{Y}}$, $Z'' = \bar{\bar{Z}}$?

12. Is this new mapping the same as the mapping found in Step 8?

In the Experiment we saw how we can create new mappings by combining two mappings which are already known.

In the first part of the Experiment, you took a point X and first found $R_L(X)$. You then rotated $R_L(X)$ by 90° around O, and thus you found the point

$$G_{90}(R_L(X)).$$

Let P be any point in the plane, and let

$$P'' = G_{90}(R_L(P))$$

as in the first part of the Experiment. The association $P \mapsto P''$ is a new mapping, which can be written

$$P \mapsto G_{90}(R_L(P)),$$

and which is called the **composition** of R_L with G_{90}.

When we combine two mappings like this, we say we are **composing** them. All we are doing is taking two mappings in succession, one after the other. The new mapping formed is called the **composite** of the two original mappings. There is a special notation for the composite mapping. In our example, the association

$$P \mapsto G_{90}(R_L(P))$$

would be denoted

$$G_{90} \circ R_L.$$

Thus $G_{90} \circ R_L$ is the new mapping formed by first doing R_L, followed by G_{90}. Thus we can also write:

$$G_{90} \circ R_L(P) = G_{90}(R_L(P)).$$

In general, if F and G are two mappings, we can form the composite mapping $F \circ G$. If P is any point in the plane, then

$$F \circ G(P) = F(G(P)).$$

Remark. The notation $F \circ G$ is read "F circle G" or "F of G". "F of G" is a good way to read the notation since it implies that we are finding the image under F *of* the image found by G. Namely, we do mapping G first, followed by F.

In Part II of the Experiment, we composed the reflection and rotation in the reverse order. We first rotated the points, and then reflected them. The composite mapping we formed in Part II is denoted $R_L \circ G_{90}$. Thus

$$\bar{X} = R_L \circ G_{90}(X) = R_L(G_{90}(X)).$$

Since X'' was not the same point as $\bar{X}$, we see that

$$G_{90} \circ R_L \neq R_L \circ G_{90}.$$

Therefore:

When composing mappings
order often makes a difference.

One particular composition of mappings is very easy to see, but is important when working formally with compositions. Let I be the identity mapping, that is

$$I(P) = P \quad \text{for all } P.$$

Then I is an isometry, and for all isometries F we have

$$I \circ F = F \circ I = F.$$

This is immediate from the definition of the identity.

To summarize some of the results suggested by the Experiment:

Composition of Reflections

Let R_L be reflection through line L. Then

$$R_L \circ R_L = I, \quad \text{the identity.}$$

In the exercises, you will consider the case where we compose two reflections through different lines.

Composition of Rotations

In Part III, we composed two rotations, to form the new mappings $G_{60} \circ G_{40}$. In Part IV, we reversed the order again, and formed the composite mapping $G_{40} \circ G_{60}$. In the case of rotations around the same point, you probably noticed that the order does **not** make a difference. We have that

$$G_{60} \circ G_{40} = G_{40} \circ G_{60}.$$

You may also have noticed that the mapping formed by composing two rotations around the same point is another rotation. In the Experiment, the composite mapping is rotation by $100°$. This is an important property, which we sum up as follows:

The composition of two rotations (with respect to the same point) is also a rotation. If x, y are numbers and G_x, G_y are rotations around the same point, then

$$G_x \circ G_y = G_{x+y} = G_y \circ G_x.$$

Example. $G_{45} \circ G_{45} = G_{90}$.

Example. $G_{180} \circ G_{270} = G_{450} = G_{90}$ as illustrated below.

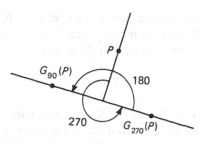

Figure 12.14

Composition of translations

If F, T are translations, then the composite $F \circ T$ is also a translation. In fact,

$$\boxed{T_A \circ T_B = T_{A+B}.}$$

Indeed, for any point P we have

$$T_A(T_B(P)) = T_A(P + B) = P + B + A = P + A + B = T_{A+B}(P).$$

This proves our assertion.

In Experiment 12-2 you should have found that the composition of isometries is again an isometry in each special case that was measured. We now prove the general fact behind this.

Theorem 12-2. *Let F and G be isometries. Then the mapping*

$$F \circ G$$

is also an isometry.

Proof. To prove that a mapping is an isometry, we have to show that the distance between two points is the same as the distance between the images of the two points under the mapping. So we choose any two points P and Q.

Since G is an isometry, we know that the distance between P and Q is equal to the distance between $G(P)$ and $G(Q)$.

Since F is an isometry, we know that the distance between $G(P)$ and $G(Q)$ is equal to the distance between $F(G(P))$ and $F(G(Q))$, which is equal to the distance between $(F \circ G)(P)$ and $(F \circ G)(Q)$.

This means that $F \circ G$ is an isometry, and proves the theorem.

We can compose more than two isometries step by step. Let F, G, H be three isometries. We can compose G and H to form the new isometry $G \circ H$. We may now compose it with F, to get

$$F \circ (G \circ H).$$

This is an isometry since F and $G \circ H$ are both isometries.

Of course we could have started by composing F and G, to get the isometry $(F \circ G)$. We then compose this isometry with H to get the new isometry $(F \circ G) \circ H$. The question now is whether this is equal to the one we formed just previously. In other words, is the following true:

$$F \circ (G \circ H) = (F \circ G) \circ H?$$

Well, we know two mappings are equal if they each map an arbitrary point into the same image point. Let P be any point in the plane. Applying our definitions of composition carefully, we have

$$F \circ (G \circ H)(P) = F((G \circ H)(P)) = F\big(G(H(P))\big)$$

and

$$(F \circ G) \circ H(P) = (F \circ G)(H(P)) = F\big(G(H(P))\big).$$

Since the right-hand side of both equations are the same, we see that the two composite isometries are the same. They both boil down to doing H, followed by G, followed by F.

Remark. Taking the composition of isometries is similar, although not exactly the same, as taking "reflections" in succession in Chapter 1, §7. They both amount to doing something, then doing something else, then still something else, successively.

Inverses of Isometries

We have seen that the composition of a reflection through a line with reflection through the same line produces the identity mapping. That is, if R_L is the reflection through a given line L, then

$$R_L \circ R_L(P) = P \qquad \text{for all points } P,$$

and therefore

$$R_L \circ R_L = I.$$

Consider the mapping G_{90}, rotation by 90° with respect to a given point O. Is there a mapping we can compose with G_{90} so that the result is the identity mapping? Again, the answer is yes:

$$G_{90} \circ G_{-90} = I.$$

In general, let F be an isometry. We define an **inverse** for F to be an isometry T such that

$$T \circ F = I \quad \text{and} \quad F \circ T = I.$$

Looking at the above examples, we would say that:

the inverse of G_{90} is G_{-90};
the inverse of R_L is R_L.

Does every isometry have an inverse? We shall answer this question in §6. It will follow from Theorem 12-12 that every isometry has an inverse. More generally, does every mapping have an inverse? The answer is no, see Exercise 17.

For the moment, we construct inverses for the standard isometries which we have been studying.

Rotations. Let G_x be rotation by x degrees, with respect to a given point O. Then G_{-x} is an inverse for G, because

$$G_{-x} \circ G_x = G_0 = I = G_x \circ G_{-x}.$$

Translations. Let T_A be translation. Then T_{-A} is an inverse for T_A because

$$T_{-A} \circ T_A = T_O = I = T_A \circ T_{-A}.$$

Reflection Through a Line. Let R be the reflection through a line L. Then R has an inverse, which is R itself, because

$$R \circ R = I.$$

In general, if F has an inverse T, then this inverse is unique.

In other words, if T_1 is another inverse, then $T_1 = T$. This is easily proved as follows:

$$T = T \circ I = T \circ F \circ T_1 = I \circ T_1 = T_1.$$

The inverse of an isometry F is denoted by F^{-1}. Thus we have the relation

$$F^{-1} \circ F = F \circ F^{-1} = I.$$

With this notation, we can write:

If G_x is a rotation, then $G_x^{-1} = G_{-x}$.

If T_A is a translation, then $T_A^{-1} = T_{-A}$.

If R is a reflection through a line, then $R^{-1} = R$.

12, §3. EXERCISES

1. In this exercise, all rotations are with respect to a single point O. For each of the following, give the single rotation G_x which is *equal* to the composition of the two given rotations, with $0 \leqq x < 360°$.
   (a) $G_{30} \circ G_{65}$ (b) $G_{180} \circ G_{180}$ (c) $G_{170} \circ G_{225}$ (d) $G_{980} \circ G_{100}$

2. Draw three points A, B, and C on your paper. What single translation is equal to each of the following:
   (a) $T_{BC} \circ T_{AB}$ (b) $T_{BA} \circ T_{AB}$ (c) $T_{CB} \circ T_{AC}$ (d) $T_{CA} \circ T_{BC}$

3. In the figure below, let R_L be reflection through line L, let R_K be reflection through line K, and let R_O be reflection through point O.

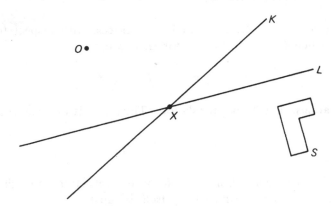

Figure 12.15

Let S be the set of points comprising the "L"-shape. Draw carefully the image of S under each of the following mappings (label each image a, b, c, or d):
   (a) R_K (b) $R_O \circ R_L$ (c) $R_L \circ R_O$ (d) $R_K \circ R_K$

4. Give examples of two isometries F and G where $F \circ G = G \circ F$.

5. Give examples of two isometries F and G where $F \circ G \neq G \circ F$.

6. What simple mapping is equal to $R_L \circ R_L$? to $R_L \circ R_L \circ R_L$?

7. What must be true about vector $\overrightarrow{AB}$ and line L so that

$$T_{AB} \circ R_L = R_L \circ T_{AB}?$$

8. Let P be the point shown below:

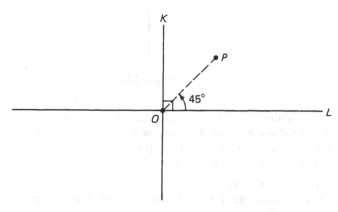

Figure 12.16

Let R_K be reflection through line K, and let R_L be reflection through line L.
The rotations G are around point O. Draw the image of P under the following isometries: (use P^a, P^b, etc. to indicate answers)
(a) R_K (b) G_{180} (c) $R_K \circ G_{90}$ (d) $R_L \circ R_K \circ G_{90}$

9. In each case name a single isometry which is equivalent to the composition mapping.
(a) $T_{BC} \circ T_{AB} = $ _____.
(b) $G_{145} \circ G_{35} = $ _____.
(c) $R_O \circ G_{30} = $ _____   (R_O is reflection through point O.)
(d) $R_O \circ R_O = $ _____.

10. Line L intersects line K at O, and $L \perp K$. In this exercise you will prove that

$$R_L \circ R_K = \text{reflection through point } O$$

in two ways.
(a) Choose a random point X. Look at X, $R_K(X)$, and $R_L \circ R_K(X)$. Using right triangles (and **RT**), show that $R_L \circ R_K(X) = $ reflection of X through O.
(b) Consider lines L and K as a pair of coordinate axes, and point O as the origin. Let $X = (x_1, x_2)$. Use the coordinate definitions of reflection through the x- and y-axis and reflection through the origin to prove the desired equality.

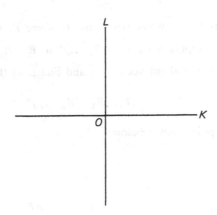

Figure 12.17

11. Let R be reflection through the origin. Let $A = (-3, 4)$.
    (a) For the following points X, draw the point $R \circ T_A \circ R(X)$
        (i) $X = (5, 6)$　　　(ii) $X = (0, 0)$　　　(iii) $X = (-1, -1)$
    (b) Does $R \circ T_A \circ R = T_A$? If not, what does it equal?

12. Let Q be the point $(3, 4)$.
    (a) If P is the point $(0, 5)$, what is the image of P when *reflected through Q* (written $R_Q(P)$)?
    (b) Same question for $P = (-1, 2)$.

13. Let T be a translation and suppose that $T = T^{-1}$. What can you say about T?

14. Let G be a rotation and suppose that $G = G^{-1}$. Is G necessarily equal to I?

15. Let F, T be isometries which have inverses F^{-1} and T^{-1} respectively. Show that $T \circ F$ has an inverse, and express this inverse in terms of F^{-1} and T^{-1}.

16. Suppose R_1, R_2, R_3 are reflections, and $F = R_1 \circ R_2 \circ R_3$. What is F^{-1}? Express F^{-1} as a composition of reflections.

17. Consider the mapping F defined by $(x, y) \mapsto (x, 0)$.
    (a) Find $F(P)$ for each point P below:

    $P = (3, 5); \quad P = (3, -2); \quad P = (-6, 0); \quad P = (-6, 10).$

    (b) Is F an isometry?
    (c) Is there an inverse of F? In other words can you define a mapping G so that $G \circ F(P) = F \circ G(P) = P$ for all points P?
    (d) At this point, what can you say about the question: Does every mapping have an inverse?

The next two sections are not needed for §6, and you may wish to read §6 before you read §4 and §5. Conversely, §6 is not needed for §4 and §5.

12, §4. DEFINITION OF CONGRUENCE

At the beginning of Chapter 7 we already dealt informally with the notion of congruence, but applied it to triangles, dealing with sides and angles. We now deal with arbitrary figures in the plane.

For instance, the quadrilaterals illustrated in Figure 12.18 are "alike" in ways similar to the two triangles at the beginning of Chapter 11, §1.

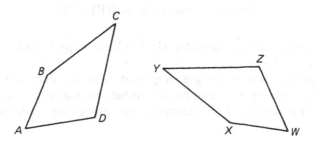

Figure 12.18

We might say that one is an exact copy of the other, and we can "lay one over the other without changing its shape", as we said in Chapter 11. We wish to make somewhat more precise what we mean by "laying one figure on another".

We say that a figure U is **congruent** to a figure V if there exists an isometry F such that the image of U under mapping F is V. In other words, if $F(U) = V$. We denote that two figures U and V are congruent by the symbols

$$U \simeq V.$$

If you think of rotations, translations, and reflections, and the application of these mappings in succession, you should see that our definition fits our intuition perfectly. It is precisely such mappings which allow us to pick one figure up, and, without distorting it, place it on top of the other figure so that the two figures match up point for point. This is what an isometry does: It "moves" figures around without changing their size or shape.

There are three useful and obvious properties of congruence:

1. *Every figure is congruent to itself*—we just choose the identity mapping as the isometry.
2. *Let S and U be two figures. If $S \simeq U$ then $U \simeq S$.*

The proof depends on the fact that if there is an isometry F such that $F(S) = U$, then there is an "inverse" isometry G such that $G(U) = S$. We shall not go here into this question of the existence of the inverse.

3. *Let U, V, W be figures in the plane. If* $U \simeq V$ *and* $V \simeq W$, *then* $U \simeq W$.

Proof. Since $U \simeq V$, there is an isometry F such that $F(U) = V$. Since $V \simeq W$, there is an isometry G such that $G(V) = W$. Then $G \circ F$ is also an isometry, and

$$G \circ F(U) = G(F(U)) = G(V) = W.$$

Since we have found an isometry $G \circ F$ which maps U onto W, we can conclude that $U \simeq W$.

To prove that two figures are congruent using the definition, we try to find an isometry, or composition of isometries usually, which will map one onto the other. For example, we prove the following simple theorem.

Theorem 12-3. *Any two segments of the same length are congruent.*

Proof. Let $\overline{PQ}$ and $\overline{MN}$ be segments of the same length.

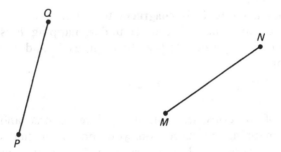

Figure 12.19

Let T be the translation which maps M onto P, so $T(M) = P$. Then

$$d(T(N), P) = d(T(N), T(M))$$

$$= d(N, M) \qquad \text{because } T \text{ preserves distances}$$

$$= d(P, Q) \qquad \text{because } |PQ| = |MN| \text{ by assumption.}$$

See Figure 12.20(a).

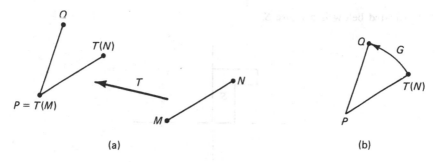

Figure 12.20

Since Q and $T(N)$ are the same distance from P, there is a rotation G around P that will map $T(N)$ onto Q (see Figure 12.20(b)).

Of course G leaves P fixed, and $G(T(N)) = Q$. Thus we have:

$$G \circ T(M) = P$$

and

$$G \circ T(N) = Q.$$

By **ISOM 1**, $G \circ T$ maps the entire segment $\overline{MN}$ onto $\overline{PQ}$, and the segments are congruent. We write:

$$\overline{MN} \simeq \overline{PQ}.$$

We could have chosen another isometry which would have mapped M onto Q and N onto P, and this would have worked as well. To show $\overline{MN} \simeq \overline{PQ}$, all we have to do is find at least one.

12, §4. EXERCISES

1. Pick out as many pairs of congruent figures as you can.

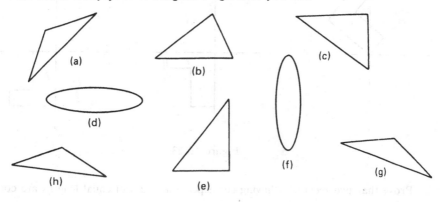

Figure 12.21

2. Illustrated below is a figure S.

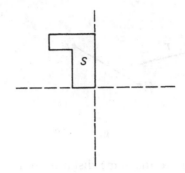

Figure 12.22

Describe what isometries you would use to map the figure into each of the positions pictured in A through E of Figure 12.23.

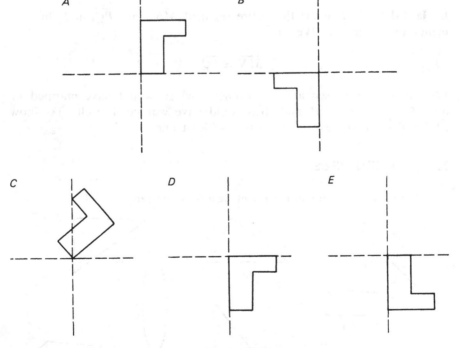

Figure 12.23

3. Prove that two rectangles having corresponding sides of equal lengths are congruent.

4. Quadrilateral $ABCD$ is a "kite", meaning that

$$|AB| = |BC|$$

and

$$|AD| = |DC|.$$

Find an isometry F of the quadrilateral with itself such that $F(A) = F(C)$. [This provides a proof that $m(\angle A) = m(\angle C)$ in a system different from Euclid's three tests. Cf. Theorem 12-4 in the next section.]

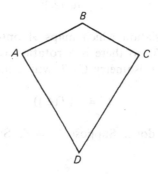

Figure 12.24

5. Prove that two circles of the same radius are congruent. [If O, O' are the centers of the circles, find an isometry mapping O on O'.]

12, §5. PROOFS OF EUCLID'S TESTS FOR CONGRUENT TRIANGLES

In this section we describe how one can prove the three criteria SSS, SAS, ASA for congruent triangles by means of isometries.

Theorem 12-4. *Let* $\triangle ABC$ *and* $\triangle XYZ$ *be triangles whose sides have the same lengths. Then the triangles are congruent.*

Proof. We suppose the sides satisfy:

$$|AB| = |XY|, \qquad |BC| = |YZ|, \qquad |AC| = |XZ|$$

as shown in Figure 12.25. We want to prove that the triangles are congruent.

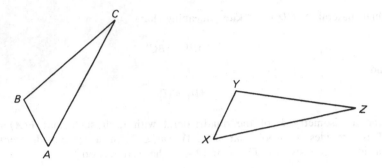

Figure 12.25

Let T be the translation which maps A onto X. Then $T(B)$ and Y are equidistant from X, so there is a rotation G such that $G(T(B)) = Y$. We have now found an isometry $G \circ T$ which maps $\overline{AB}$ onto $\overline{XY}$. Let

$$C' = G(T(C)).$$

If $C' = Z$ then we are done. Suppose $C' \neq Z$. Since $G \circ T$ is an isometry, we conclude that

$$|XZ| = |XC'| \qquad \text{and} \qquad |YZ| = |YC'|.$$

By the perpendicular bisector Theorem 5-1, we conclude that X, Y are on the perpendicular bisector of $\overline{ZC'}$.

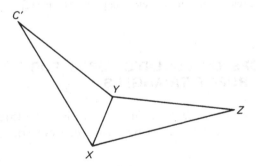

Figure 12.26

In that case, let R be reflection through L_{XY}. Then R maps C' on Z. Thus the isometry $R \circ G \circ T$ maps ABC onto XYZ, and we have

$$\triangle ABC \simeq \triangle XYZ.$$

Theorem 12-5. *If two triangles have one corresponding side of the same length and two corresponding angles of the same measure, then the triangles are congruent.*

Proof. Suppose first that the corresponding side is a common side for the two angles. Given $\triangle ABC$ and $\triangle XYZ$, with $|AB| = |XY|$. $m(\angle A) = m(\angle X)$ and $m(\angle B) = m(\angle Y)$, as shown:

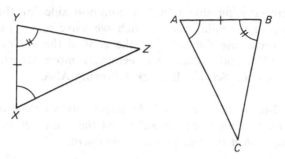

Figure 12.27

There exists a translation T which maps A onto X. Since $|XY| = |AB|$, there is a rotation G around X which will then map $T(B)$ onto Y. As before, we have simply mapped segment $\overline{AB}$ onto $\overline{XY}$. Let C' be the image of C under the composition $G \circ T$. Two cases may arise:

C' may lie on the same side of segment $\overline{XY}$ as Z does:

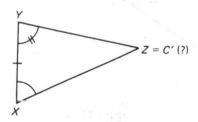

Figure 12.28

Since $m(\angle A) = m(\angle X)$, the image of ray R_{AC} must coincide with ray R_{XZ}. Similarly, since $m(\angle B) = m(\angle Y)$, the image of ray R_{BC} must coincide with ray R_{YZ}. Thus $C' = Z$, and the triangles are congruent.

The other possibility is that C' may lie on the other side of $\overline{XY}$, as illustrated:

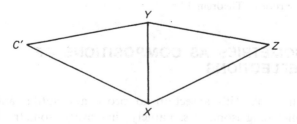

Figure 12.29

In this case, we reflect the triangle XYC' through line L_{XY}, and we reduce the problem to the situation just discussed. This completes the proof in the present case.

If the corresponding side is not a common side for the two angles, then we can give a similar proof which we leave as an exercise for the reader. Or we can use the fact that the sum of the angles of a triangle has measure 180°, and reduce this seemingly more general case to the case already treated. See the Remark following **ASA**.

Theorem 12-6. *If two sides and the angle between them in one triangle have the same measures as two sides and the angle between them in the other triangle, then the triangles are congruent.*

Proof. Given $\triangle ABC$ and $\triangle XYZ$, with $|AB| = |XY|$, $|AC| = |XZ|$, and $m(\angle A) = m(\angle X)$:

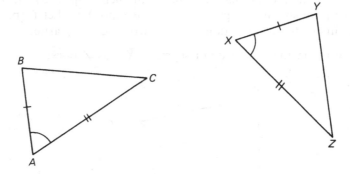

Figure 12.30

The details of this proof are left as an exercise. Proceed in a manner similar to Theorems 12-4 and 12-5.

12, §5. EXERCISE

1. Finish the proof of Theorem 12-6.

12, §6. ISOMETRIES AS COMPOSITIONS OF REFLECTIONS

We end our book with a section to prove a beautiful and important theorem concerning isometries, namely that every isometry is either the identity or can be expressed as the composition of at most three reflections. In other words, all isometries are, fundamentally, reflections or compositions of reflections. The beauty and importance of this result is that it reduces our motley (and perhaps incomplete) collection of reflections, translations, and rotations, to a single, all-inclusive concept. This kind of result not only pleases one's mathematical sense, but also allows us to discover properties which may not have been apparent before. For example, it will allow us to answer a key question, namely that every isometry does indeed have an inverse.

Not only mathematicians appreciate such simplifications. Much of the work of physicists is aimed toward the same goal: classify the symmetries of nature, and reduce fundamental concepts to as short a list as possible.

Before we can classify isometries, however, we need some preparatory material.

We recall that a **fixed point** P for an isometry F is a point such that

$$F(P) = P.$$

Look at Exercises 5, 6, 7, 8 of §1, as preparation for what comes next.

Theorem 12-7. *Let F be an isometry, and suppose F has two distinct fixed points P and Q. Then every point of the line L_{PQ} is a fixed point.*

Proof. Let X be a point on L_{PQ}. Suppose first that X lies on the segment $\overline{PQ}$. Then

$$d(P, Q) = d(P, X) + d(X, Q).$$

Since F is an isometry, we have

$$d(P, X) = d(F(P), F(X)) = d(P, F(X))$$

and

$$d(X, Q) = d(F(X), F(Q)) = d(F(X), Q).$$

Hence

$$d(P, Q) = d(P, F(X)) + d(F(X), Q).$$

Use the results of the Exercises mentioned above and postulate **SEG** to show that $F(X) = X$. Use a similar argument for the case where X lies on L_{PQ} but not on segment $\overline{PQ}$ to conclude the proof.

Fixed points are the stepping stones to proving our desired result. The approach is to characterize an isometry by the number of its fixed points, and to investigate the results of composing it with reflections.

First, in Theorem 12-8, we show that if an isometry leaves fixed three points which are not on the same line, then that isometry must be the identity. This is not a surprising result, and it can be proved fairly easily. The importance of it, however, is that it reduces our problem to considering only those isometries which leave zero, one, or two points fixed! An isometry with three or more fixed points is necessarily the identity.

In Theorem 12-9 we place the first stone and consider the first case. We show that an arbitrary isometry (with perhaps **zero** fixed points) must either be the identity, or it can be composed with a reflection so that the resulting composition mapping has one fixed point.

Then, in Theorem 12-10, we place the next stone. We show that an isometry with **one** fixed point must either have a second fixed point as well, or it can be composed with a reflection so that the resulting composition mapping has two fixed points.

Theorem 12-11 places the last stone. An isometry with **two** fixed points must either have a third fixed point or it can be composed with a reflection so that the resulting composition mapping has three fixed points. In other words, an isometry with two fixed points is either the identity (has three fixed points to start with) or is a reflection (since composing it with a reflection produces the identity).

Our desired Theorem, 12-12, forces any isometry to walk the path. It concludes that an isometry is the identity or a composition of at most three reflections. You'll see.

Theorem 12-8. *Let F be an isometry. If F has three distinct fixed points, not lying on the same line, then F is the identity.*

Proof. Let P, Q, M be the three fixed points. By Theorem 12-7 we know that the lines L_{PQ} and L_{QM} are fixed. Let X be any point. If X is on L_{PQ} or L_{QM} then X is fixed by Theorem 12-7, and we are done, so we suppose X is not on the lines L_{PQ} or L_{QM}. Let L be a line through X which is not parallel to L_{PQ} or L_{QM}, and also is not equal to L_{PX} or L_{QX} or L_{MX}.

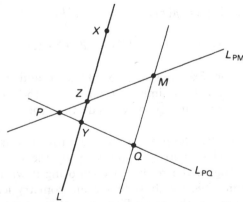

Figure 12.31

Thus we exclude only a finite number of lines. Then L intersects L_{PQ} in a point $Y \neq P$, and also L intersects L_{PM} in a point $Z \neq M$. Furthermore $Y \neq Z$, otherwise L_{PQ} and L_{PM} contain a common point $Y = Z$, but $Y \neq P$, $Y \neq Q$, and $Y \neq M$. This would imply that $L_{PQ} = L_{QM}$, which contradicts our assumption that P, Q, M are not on the same line. By Theorem 12-7 we deduce that L_{YZ} is fixed, and hence X is a fixed point, thus concluding the proof.

Theorem 12-9. *Let F be an isometry. Then either F is the identity, or there exists a reflection R such that $R \circ F$ has one fixed point.*

Proof. Suppose F is not the identity. Let P be a point such that $F(P) \neq P$. Let $F(P) = P'$. Let L be the perpendicular bisector of the segment $\overline{PP'}$. Let R be the reflection through L. Then

$$R(P') = P,$$

and so

$$R(F(P)) = P.$$

Hence $R \circ F$ has a fixed point, thus proving the theorem.

Theorem 12-10. *Let F be an isometry with one fixed point. Then F has two fixed points or there exists a reflection R such that $R \circ F$ has two fixed points.*

Proof. Let P be the fixed point of F, so $F(P) = P$. Let X be any other point, and let $F(X) = X'$.

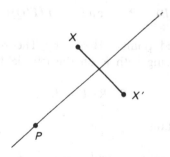

Figure 12.32

If $X = X'$ then we are done. Suppose $X \neq X'$. Since F is an isometry, we have

$$d(P, X) = d(F(P), F(X)) = d(P, F(X)).$$

By Theorem 5-1 of Chapter 5, §1 we conclude that P lies on the perpendicular bisector of the segment between X and $F(X)$. Hence X' is the reflection of X through this bisector. Let R be this reflection. Then $R(X') = X$, and so

$$R(F(X)) = X.$$

But $R(F(P)) = R(P) = P$, so $R \circ F$ has two fixed points, as was to be shown.

Theorem 12-11. *Let F be an isometry which has two distinct fixed points P and Q. Then either F is the identity, or F is the reflection through the line L_{PQ}.*

Proof. Suppose F is not the identity. By Theorem 12-7 we know that F leaves every point on L_{PQ} fixed. Let X be a point not on L_{PQ} such that $F(X) \neq X$. Since F is an isometry, we have

$$d(P, X) = d(F(P), F(X)) = d(P, F(X)).$$

By Theorem 5-1 of Chapter 5, §1 we conclude that P lies on the perpendicular bisector of the segment between X and $F(X)$. Similarly, Q lies on this perpendicular bisector. Therefore the line L_{PQ} is the perpendicular bisector of this segment. Hence $F(X)$ is the reflection of X through L_{PQ}. Let R be this reflection. Then

$$R(F(X)) = X.$$

But

$$R(F(P)) = R(P) = P \quad \text{and} \quad R(F(Q)) = R(Q) = Q,$$

so $R \circ F$ has three fixed points. Hence by Theorem 12-8, we conclude that $R \circ F = I$. Composing with R on the left yields

$$R \circ R \circ F = R.$$

Since $R \circ R = I$, we obtain

$$I \circ F = R \quad \text{and hence} \quad F = R,$$

thereby concluding the proof.

Finally, we are able to prove the main theorem of this section.

Theorem 12-12. *Let F be an isometry. Then F is the identity, or F is a composition of at most three reflections.*

Proof. If F is the identity, we are done. If not, then by Theorem 12-9 there is a reflection R_1 such that $R_1 \circ F$ has a fixed point. If $R_1 \circ F = I$, then composing with R_1 on the left yields

$$R_1 \circ R_1 \circ F = R_1 \qquad \text{and hence} \qquad F = R_1$$

because $R_1 \circ R_1 = I$ and $I \circ F = F$. We are done. Suppose $R_1 \circ F \neq I$. By Theorem 12-10 there exists a reflection R_2 such that $R_2 \circ R_1 \circ F$ has two fixed points. If $R_2 \circ R_1 \circ F = I$ then we compose with $R_1 \circ R_2$ on the left and we get

$$R_1 \circ R_2 \circ R_2 \circ R_1 \circ F = R_1 \circ R_2 \qquad \text{and hence} \qquad F = R_1 \circ R_2$$

because $R_2 \circ R_2 = I$ and $R_1 \circ R_1 = I$ and $I \circ F = F$. We are done. Suppose that $R_2 \circ R_1 \circ F \neq I$. By Theorem 12-11 there exists a reflection R_3 such that

$$R_2 \circ R_1 \circ F = R_3.$$

Then we compose with $R_1 \circ R_2$ on the left, and we obtain

$$F = R_1 \circ R_2 \circ R_3.$$

This concludes the proof.

12, §6. EXERCISES

1. Finish the proof of Theorem 12-7.

2. Let F, T be isometries. Let P, Q, M be three points not on a straight line such that

$$F(P) = T(P), \qquad F(Q) = T(Q), \qquad F(M) = T(M).$$

   Prove that $F = T$.

3. Let F be an isometry with one fixed point. Prove that there is a rotation G such that $G \circ F$ has two fixed points.

4. Let F be an isometry. Prove that there is a translation T such that $T \circ F$ has a fixed point.

5. Prove that every isometry F can be written in the form

$$F = T \circ G \circ R \qquad \text{or} \qquad F = T \circ G, \qquad \text{or} \qquad F = T \qquad \text{or} \qquad F = I,$$

   where R is a reflection through a line, G is a rotation, and T is a translation.

6. Prove that every isometry has an inverse. [See Exercise 16 in §3.]

Index